NUCLEAR POWER PLANTS

DESIGN AND SAFETY CONSIDERATIONS

NUCLEAR MATERIALS AND DISASTER RESEARCH

Additional books in this series can be found on Nova's website under the Series tab.

Additional E-books in this series can be found on Nova's website under the E-books tab.

SAFETY AND RISK IN SOCIETY

Additional books in this series can be found on Nova's website under the Series tab.

Additional E-books in this series can be found on Nova's website under the E-books tab.

NUCLEAR MATERIALS AND DISASTER RESEARCH

NUCLEAR POWER PLANTS

DESIGN AND SAFETY CONSIDERATIONS

JAMES P. ARGYRIOU
EDITOR

Nova Science Publishers, Inc.
New York

Copyright ©2012 by Nova Science Publishers, Inc.

For permission to use material from this book please contact us:
Telephone 631-231-7269; Fax 631-231-8175
Web Site: http://www.novapublishers.com

NOTICE TO THE READER

The Publisher has taken reasonable care in the preparation of this book, but makes no expressed or implied warranty of any kind and assumes no responsibility for any errors or omissions. No liability is assumed for incidental or consequential damages in connection with or arising out of information contained in this book. The Publisher shall not be liable for any special, consequential, or exemplary damages resulting, in whole or in part, from the readers' use of, or reliance upon, this material. Any parts of this book based on government reports are so indicated and copyright is claimed for those parts to the extent applicable to compilations of such works.

Independent verification should be sought for any data, advice or recommendations contained in this book. In addition, no responsibility is assumed by the publisher for any injury and/or damage to persons or property arising from any methods, products, instructions, ideas or otherwise contained in this publication.

This publication is designed to provide accurate and authoritative information with regard to the subject matter covered herein. It is sold with the clear understanding that the Publisher is not engaged in rendering legal or any other professional services. If legal or any other expert assistance is required, the services of a competent person should be sought. FROM A DECLARATION OF PARTICIPANTS JOINTLY ADOPTED BY A COMMITTEE OF THE AMERICAN BAR ASSOCIATION AND A COMMITTEE OF PUBLISHERS.

Additional color graphics may be available in the e-book version of this book.

Library of Congress Cataloging-in-Publication Data

Nuclear power plants : design and safety considerations / editor, James P. Argyrio.
    p. cm.
Includes bibliographical references and index.
ISBN 978-1-61470-952-7 (softcover)
1. Nuclear power plants--United States--Safety measures. 2. Nuclear power plants--United States--Design and construction. I. Argyriou, James P.
TK1343.N75 2011
621.48'3--dc23
                                    2011028967

Published by Nova Science Publishers, Inc. † *New York*

CONTENTS

PREFACE

The seismic design criteria applied to sitting commercial nuclear power plants operating in the U.S. received increased attention following the March 11th, 2011 earthquake and tsunami that devastated Japan's Fukushima Daiichi nuclear power station. Since the events, some in Congress have begun to question whether U.S. plants are vulnerable to a similar threat, particularly in light of the Nuclear Regulatory Commissions's (NRC)ongoing reassessment of seismic risks at certain plant sites. This book presents some of the general design concepts of operating nuclear power plants in order to discuss design considerations for seismic events.

Chapter 1- Since the March 11, 2011, earthquake and tsunami that devastated Japan's Fukushima Daiichi nuclear power station, the seismic criteria applied to siting commercial nuclear power plants operating in the United States have received increased attention; particularly the Nuclear Regulatory Commission's (NRC's) 2010 reassessment of seismic risks at certain plant sites.

Chapter 2- Nuclear power plants are built to withstand environmental hazards, including earthquakes. Even those plants that are located outside of areas with extensive seismic activity are designed for safety in the event of such a natural disaster. The Nuclear Regulatory Commission (NRC) requires all of its licensees to take seismic activity into account when designing and maintaining its nuclear power plants. When new seismic hazard information becomes available, the NRC evaluates the new data and models and determines if any changes are needed at plants. The newest seismic data suggests that although the potential seismic hazard at some nuclear power plants in central and eastern states may have increased beyond previous estimates, all operating nuclear plants remain safe with no need for immediate action.

Chapter 3- The NRC has long sought standardization of nuclear power plant designs, and the enhanced safety and licensing reform that standardization could make possible. The Commission expects advanced reactors to be safer and use simplified, passive or other innovative means to accomplish their safety functions. The NRC's regulation (Part 52 to Title 10 of the Code of Federal Regulations) provides a predictable licensing process including certification of new nuclear plant designs. This process reflects decades of experience and research involving reactor design and operation. The design certification process provides for early public participation and resolution of safety issues prior to an application to construct a nuclear power plant.

Chapter 4- In May 2009, President Barack Obama called for harnessing the power of nuclear energy "on behalf of our efforts to combat climate change and to advance peace and opportunity for all people." Meeting the energy, environment, and climate demands of the 21st century will require creating new solutions and reimagining older but still crucial technologies. Civil nuclear technology combines elements of both approaches.

Chapter 5- The physical security of nuclear power plants and their vulnerability to deliberate acts of terrorism was elevated to a national security concern following the attacks of September 11, 2001. Since the attacks, Congress has repeatedly focused oversight and legislative attention on nuclear power plant security requirements established and enforced by the Nuclear Regulatory Commission (NRC

Chapter 6- The Price-Anderson Act, which became law on September 2, 1957, was designed to ensure that adequate funds would be available to satisfy liability claims of members of the public for personal injury and property damage in the event of a nuclear accident involving a commercial nuclear power plant. The legislation helped encourage private investment in commercial nuclear power by placing a cap, or ceiling on the total amount of liability each holder of a nuclear power plant licensee faced in the event of an accident. Over the years, the "limit of liability" for a nuclear accident has increased the insurance pool to more than $12 billion.

In: Nuclear Power Plants   ISBN: 978-1-61470-952-7
Editor: James P. Argyriou   ©2012 Nova Science Publishers, Inc.

Chapter 1

NUCLEAR POWER PLANT DESIGN AND SEISMIC SAFETY CONSIDERATIONS

Anthony Andrews

SUMMARY

Since the March 11, 2011, earthquake and tsunami that devastated Japan's Fukushima Daiichi nuclear power station, the seismic criteria applied to siting commercial nuclear power plants operating in the United States have received increased attention; particularly the Nuclear Regulatory Commission's (NRC's) 2010 reassessment of seismic risks at certain plant sites.

Commercial nuclear power plants operating in the United States vary considerably, as most were custom-designed and custom-built. Boiling water reactors (BWRs) directly generate steam inside the reactor vessel. Pressurized water reactors (PWRs) use heat exchangers to convert the heat generated by the reactor core into steam outside of the reactor vessel. U.S. utilities currently operate 104 nuclear power reactors at 65 sites in 31 states; 69 are PWR designs and the 35 remaining are BWR designs.

One of the most severe operating conditions for a reactor is a loss of coolant accident (LOCA), which can lead to a reactor core meltdown. The emergency core cooling system (ECCS) provides core cooling to minimize fuel damage by injecting large amounts of cool, borated water into the reactor coolant system following a pipe rupture or other water loss, and (secondarily) to provide extra neutron poisons to ensure the reactor remains shut down. The ECCS must be sized to provide adequate make-up water to compensate for a break of the largest diameter pipe in the primary system (i.e., the so-called "double-ended guillotine break" (DEGB)). However, the NRC considers the DEGB to be an extremely unlikely event. Nevertheless, even unlikely events can occur, as the combined tsunami and magnitude 9.0 earthquake that struck Fukushima Daiichi proves.

U.S. nuclear power plants have designs based on Deterministic Seismic Hazard Analysis (DSHA). Since then, Probabilistic Seismic Hazard Analysis (PSHA) has been adopted as a more comprehensive approach in engineering practice. Consequently, the NRC is reassessing the probability of seismic core damage at existing plants.

In 2008, the U.S Geological Survey (USGS) released an update of the National Seismic Hazard Maps (NSHM). USGS notes that the 2008 hazard maps differ significantly from the 2002 maps in many parts of the United States, and generally show 10%-15% reductions in spectral and peak ground acceleration across much of the Central and Eastern United States (CEUS), and about 10% reductions for spectral and peak horizontal ground acceleration in the Western United States (WUS). Seismic hazards are greatest in the WUS, particularly in California, Oregon, and Washington, as well as Alaska and Hawaii.

In 2010, NRC published its GI-199 Safety/Risk Assessment; a two-stage assessment of the implications of USGS updated probabilistic seismic hazards analysis in the CEUS on existing nuclear power plants sites. NRC does not rank nuclear plants by seismic risk. NRC's objective in GI-199 was to evaluate the need for further investigations of seismic safety for operating reactors in the CEUS. The data evaluated in the assessment suggest that the probability for earthquake ground motion above the seismic design basis for some nuclear plants in the CEUS, although still low, is larger than previous estimates. In late March 2011, NRC announced that it had identified 27 nuclear reactors operating in the CEUS that would receive priority earthquake safety reviews.

BACKGROUND ON SEISMIC STANDARDS

The seismic design criteria applied to siting commercial nuclear power plants operating in the United States received increased attention following the March 11 earthquake and tsunami that devastated Japan's Fukushima Daiichi nuclear power station. Since the events, some in Congress have begun to question whether U.S plants are vulnerable to a similar threat, particularly in light of the Nuclear Regulatory Commission's (NRC's) ongoing reassessment of seismic risks at certain plant sites.[1]

Commercial nuclear power plants operating in the United States use light water reactor designs, but vary widely in design and construction. Light water reactors use ordinary water as a neutron moderator and coolant, and uranium fuel enriched in fissile uranium-235.[2] Designs fall into either pressurized water reactor (PWR) or boiling water reactor (BWR) categories. Both have reactor cores (the source of heat) consisting of arrays of uranium fuel bundles capable of sustaining a controlled nuclear reaction.[3] U.S. commercial nuclear power plants incorporate safety features intended to ensure that, in the event of an earthquake, the reactor core would remain cooled, the reactor containment would remain intact, and radioactive releases would not occur from spent fuel storage pools. The Nuclear Regulatory Commission (NRC) defines this as the "safe-shutdown condition."

When utilities began building nuclear power plants in the 1960s-1970s era, they typically hired an architect/engineering firm, then contracted with a reactor manufacturer ("nuclear vendors") to build the nuclear steam supply system (NSSS), consisting of the nuclear core, reactor vessel, steam generators and pressurizer (in PWRs), and control mechanisms—representing about 10% of the plant investment.[4] The balance of the plant (BOP) consisted of secondary cooling systems, feed-water systems, steam systems, control room, and generator systems. At the time, the four vendors who offered designs for nuclear reactor systems were Babcock & Wilcox, Combustion Engineering, General Electric, and Westinghouse. About 12 architect/engineering firms were available to design the balance of the plant. Each architect/engineer had its own preferred approach to designing the balance of plant systems. In addition, plant site-conditions varied due to the different meteorological, seismic, and hydrological conditions. The custom design-and-build industry approach resulted in problems verifying the safety of individual plants and in transferring the safety lessons learned from one reactor to another.

The previous design approach to withstanding earthquakes had relied on Deterministic Seismic Hazard Analysis (DSHA). Any new plant design is to consider Probabilistic Seismic Hazard Analysis (PSHA), which has been widely adopted in engineering practice. Deterministic analysis attempts to quantify the worst-case scenario based on the combination of earthquake sources at a site's location that results in the strongest ground-motion potentially generated.[5] In other words, the deterministic assessment focuses on a single earthquake event to determine the finite probability of occurrence.

PSHA is a methodology that estimates the likelihood that various levels of earthquake-caused ground motion will be exceeded at a given location in a given future time period.[6] Due to possible uncertainties in geoscience data and in the models used to estimate ground motion from earthquakes, multiple model interpretations are often possible. This has led to disagreement among experts, which in turn has led to disagreement on the selection of ground motion magnitudes for the design at a given site. PSHA traditionally quantified ground motion based on peak ground acceleration (PGA).[7] Today, the preferred parameter is Response Spectral Acceleration (SA), which gives the maximum acceleration of an oscillating structure such as a building or power plant.

In its 2010 study (GI-199), the NRC concludes that deterministic assessments (DSHA) do not necessarily mean that the seismic design basis for the Safe Shutdown Earthquake (SSE) condition was, or is, deficient in some fashion.[8] The design approach to developing loadings on power plant piping and equipment systems relies on the SSE condition. Existing nuclear plants designs include considerable safety margins that enable them to withstand "deterministic" or "scenario earthquake" ground motions that accounted for the largest earthquakes expected in the area around the plant.[9] The NRC study found that some plant sites might have an increased probability, albeit relatively small, of exceeding their design basis ground motion. NRC considers that the probabilities of seismic core damage occurring are lower than its guidelines for taking immediate action, but has determined that some plants' performance should be reassessed based on updated seismic hazards.

This report presents some of the general design concepts of operating nuclear power plants in order to discuss design considerations for seismic events. This report does not attempt to conclude whether one design is inherently safer or less safe than another plant. Nor does it attempt to conclude whether operating nuclear power plants are at any greater or lesser risk from earthquakes given recent updates to seismic data and seismic hazard maps.

NUCLEAR POWER PLANT DESIGNS

Currently, 104 nuclear power plants currently operate at 65 sites in 31 states; 69 are PWR designs and the 35 remaining are BWR designs.

The more numerous PWR plants include Babcock & Wilcox, Combustion Engineering, and Westinghouse designs. The BWR plants all use a General

Electric design. Table 1 summarizes the various reactor types. The sections that follow discuss them further.

Table 1. Reactor Type, Vendor, and Containment.

Reactor Type	Vendor	Containment Type	No. of Plants
PWR	Babcock & Wilcox 2-Loop Lower	Dry, Ambient Pressure	7
	Combustion Engineering	Dry, Ambient Pressure	11
	Combustion Engineering System 80	Large Dry, Ambient Pressure	3
	Westinghouse 2-Loop	Dry, Ambient Pressure	6
	Westinghouse 3-Loop	Dry, Ambient Pressure	7
	Westinghouse 3-Loop	Dry, Sub-atmospheric	6
	Westinghouse 4-Loop	Dry, Ambient Pressure	18
	Westinghouse 4-Loop	Dry, Sub-atmospheric	1
	Westinghouse 4-Loop	Wet, Ice Condenser	9
	Westinghouse 4-Loop	Dry, Ambient Pressure	1
			69
BWR	General Electric Type 2	Wet, Mark I	2
	General Electric Type 3	Wet, Mark I	6
	General Electric Type 4	Wet, Mark 1	15
	General Electric Type 4	Wet, Mark II	4
	General Electric Type 5	Wet, Mark II	4
	General Electric Type 6	Wet, Mark III	4
			35

Source: U.S. NRC.

Boiling Water Reactor (BWR) Systems

A boiling water reactor generates steam directly inside the reactor vessel as water flows upward through the reactor's core (see Figure 1).[10] The water also cools the reactor core, and the reactor operator is able to vary the reactor's power by controlling the rate of water flow through the core with recirculation pumps and jet pumps. The generated steam flows out the top of the reactor vessel through pipelines to a combined high-pressure/low-pressure turbine-generator. After the exhausted steam leaves the low-pressure turbine, it runs through a condenser/heat exchanger that cools the steam and condenses it back to water. A series of pumps return the condensed water back to the reactor vessel. The heat exchanger cycles cooling water through a cooling tower, or takes in and discharges water with a lake, river, or ocean. The water that flows through the reactor, steam turbines, and condenser is a closed loop that never contacts the outside environment under normal operating conditions. Reactors of this design operate at temperatures of approximately 570° F and pressures of 1,000 pounds per square inch (psi) atmospheric.

Safe-Shutdown Condition

During normal operation, reactor cooling relies on the water that enters the reactor vessel and the generated steam that leaves. During safe shutdown, the core continues to generate heat by radioactive decay and generates steam.[11] Under this condition, the steam bypasses the turbine and diverts directly to the condenser to cool the reactor. When the reactor vessel pressure decreases to approximately 50 psi, the shutdown-cooling mode removes residual heat by pumping water from the reactor recirculation loop through a heat exchanger and back to the reactor via the recirculation loop. The recirculation loop design limits the number of pipes that penetrate the reactor vessel.

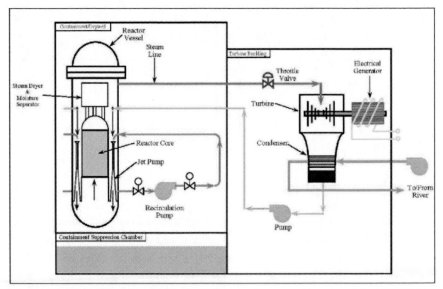

Source: U.S. Nuclear Regulatory Commission, *Reactor Concepts Manual, Boiling Water Reactor Systems, 2005.*

Figure 1. Boiling Water Reactor (BWR) Plant Generic Design Features.

Table 2. BWR Design Evolution.

Model	Year Introduced	Design Feature	Typical Plants
BWR/1	1955	Natural circulation	Dresden 1
		First internal steam separation	Big Rock Point
		Isolation condenser	Humboldt Bay
		Pressure Suppression Containment	
BWR/2	1963	Large direct cycle	Oyster Creek
BWR/3/4	1965/1966	First jet pump application	Dresden 2
		Improved Emergency Core Cooling System (ECCS); spray and flood	Browns Ferry
		Reactor Core Isolation Cooling, (RCIC) system	
BWR/5	1969	Improved ECCS systems	LaSalle
		Valve recirculation flow control	9 Mile Point 2
BWR/6	1972	Improved jet pumps and steam separators	Clinton
		Reduced fuel duty: 13.4 kW/ft, 44 kW/m	Grand Gulf
		Improved ECCS performance	Perry
		Gravity containment flooder	
		Solid-state nuclear system protection system (Option, Clinton only)	
		Compact control room option	

Source: M. Ragheb, Chapter 3, *Boiling Water Reactors*, https://netfiles.uiuc.edu/ mragheb/www/NPRE%20402%20ME%20405%20Nuclear%20Power%20Engine ering/Boiling%20Water%20Reactors.pdf.

Note: All BWR/1 plants that operated in the United States have been decommissioned.

Loss of Coolant Accident

The most severe operating condition that a reactor design must contend with is a loss of coolant accident (LOCA). In the absence of coolant, the uncovered reactor core continues to generate heat through fission. The resulting heat buildup can damage the fuel or fuel cladding and lead to a fuel "meltdown." Under such a condition, an emergency core cooling system (ECCS) provides water to cool the reactor core. The ECCS is an independent high-pressure coolant injection system that requires no auxiliary electrical power, plant air systems, or external cooling water systems to provide makeup water under small and intermediate loss of coolant accidents. A low-pressure ECCS sprays water from the suppression pool into the reactor vessel and on top of the fuel assemblies.[12] The ECCS must also be sized to provide adequate makeup water to compensate for a break of the largest diameter pipe in the primary system (i.e., the so-called "double-ended guillotine break" (DEGB)). However, the NRC views the DEGB as an extremely unlikely event (likely to occur only once per 100,000 years of reactor operation).[13]

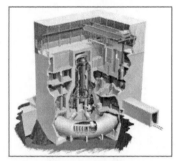

Source: General Electric, in *NRC Boiling Water Reactor (BWR) Systems*, http://www. nrc.gov/reading-rm/basic-ref/ teachers/03.pdf.
Note: Japan's Fukushima Daiichi plants use designs similar to this.

Figure 2. GE BWR / Mark I Containment Structure Showing Torus Suppression Pool.

Source: General Electric, in NRC *Boiling Water Reactor (BWR) Systems*, http://www. nrc.gov/reading-rm/basic-ref/ teachers/03.pdf.

Figure 3. General Electric Mark II Containment Structure.

Source: General Electric, in *NRC Boiling Water Reactor (BWR) Systems*, http://www.nrc.gov/reading-rm/basic-ref/ teachers/03.pdf.

Notes:

Reactor Building	Auxiliary Building	Fuel Building
1. Shield Building	16. Steam Line Channel	19. Spent Fuel Shipping cask
2. Free Standing Steel Containment	17. RHR System	20. Fuel Storage Pool
3. Upper Pool	18. Electrical Equipment Room	21. Fuel Transfer Pool
4. Refueling Platform		22. Cask Loading Pool
5. Reactor Water Cleanup		23. Cask Handling Crane
6. Reactor Vessel		24. Fuel Transfer Bridge
7. Steam Line		25. Fuel Cask Skid on Railroad Car
8. Feed-water Line		
9. Recirculation Loop		
10. Suppression Pool		
11. Weir Wall		
12. Horizontal Vent		
13. Dry Well		
14. Shield Wall		
15. Polar Crane		

Figure 4. General Electric Mark III Containment Structure.

BWR Design Evolution

Currently, General Electric Type 2 through Type 6 BWRs operate in the United States (Table 1). BWRs are inherently simpler designs than other light water reactor types. Since they heat water and generate steam directly inside the reactor vessel, there are fewer components.

Pressurized Water Reactor Systems

A pressurized water reactor (PWR) generates steam outside the reactor vessel, unlike a BWR design. A primary system (reactor cooling system) cycles superheated water from the core to a heat exchanger/steam generator. A secondary system then transfers steam to a combined high-pressure/ low-pressure turbine generator (Figure 5).[14] Steam exhausted from the low-pressure turbine runs through a condenser that cools and condenses it back to water. Pumps return the cooled water back to the steam generator for reuse. The condenser cools the steam leaving the turbine-generator through a third system by flowing past a heat-exchanger that recycles cooling water through a cooling tower, or takes in and discharges water with a lake, river, or ocean. Unlike a BWR design, the cooling water that flows through the reactor core never contacts the turbine-generator. Nor does reactor cooling water contact the environment under normal operating conditions.

To keep the reactor operating under ideal conditions, a pressurizer keeps water and steam pressure under equilibrium conditions. The pressurizer is part of the reactor coolant system, and consists of electrical heaters, pressure sprays, power-operated relief valves, and safety valves. For example, if pressure rises too high, water spray cools the steam in the pressurizer; or if pressure is too low, the heaters increase steam pressure. The cause of the pressure deviation is normally associated with a change in the temperature of the reactor coolant system.

PWR Design Configurations

All PWR systems consist of the same major components, but arranged and designed differently. For example, Westinghouse has built plants with two, three, or four primary coolant loops, depending upon the power output of the plant.

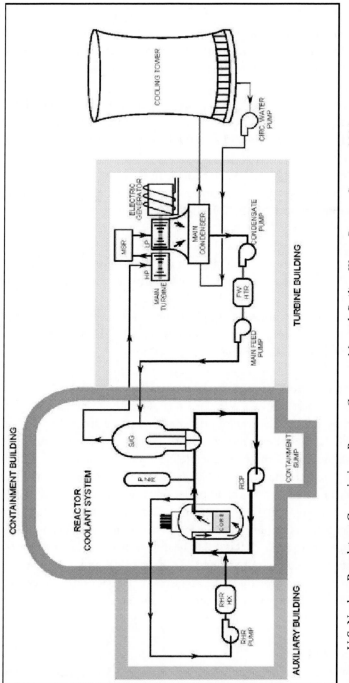

Source: U.S. Nuclear Regulatory Commission, *Reactor Concepts Manual, Boiling Water Reactor Systems*, 2005.

Notes: PIZ – Pressurizer; S/G – Steam generator

Figure 5. Pressurized Water Reactor (PWR) Plant Generic Design Features.

- Two-loop Westinghouse reactors have two steam generators, two reactor coolant pumps, a pressurizer, and 121 fuel assemblies; electrical output is approximately 500 megawatts. Six currently operate.[15]
- Three-loop Westinghouse reactors have three steam generators, three reactor coolant pumps, a pressurizer, and 157 fuel assemblies; output ranges from 700 to more than 900 megawatts. Thirteen currently operate.[16]
- Four-loop Westinghouse reactors have four steam generators, four reactor coolant pumps, a pressurizer, and 193 fuel assemblies; output ranges from 950 to 1,250 megawatts.[17] Twenty-nine currently operate.

The seven operating Babcock & Wilcox reactors have two once-through steam generators, four reactor coolant pumps, and a pressurizer.[18] These reactors have 177 fuel assemblies and produce approximately 850 megawatts of electricity.

The 14 operating Combustion Engineering reactors have two steam generators, four reactor coolant pumps, and a pressurizer.[19] They produce from less than 500 to more than 1,200 megawatts.

Safe Shutdown Condition

During normal operation, a PWR does not generate steam directly. For cooling, it transfers heat via the reactor primary coolant to a secondary coolant in the steam generators. There, the secondary coolant water is boiled into steam and sent to the main turbine to generate electricity. Even after shutdown (when the moderated uranium fission is halted), the reactor continues to produce a significant amount of heat from decay of uranium fission products (decay heat). The decay heat is sufficient to cause fuel damage if the core cooling is inadequate. Auxiliary feed-water systems and the steam dump systems work together to remove the decay heat from the reactor. If a system for dumping built-up steam is not available or inoperative, atmospheric relief valves can dump the steam directly to the atmosphere. Under normal operating conditions, water flowing through the secondary system does not contact the reactor core; dumped-steam does not present a radiological release.

Loss of Coolant Accident

The most severe operating condition that reactor designs must contend with is the loss of coolant accident (LOCA); the extreme case represented by the double-ended guillotine break (DEGB) of large diameter pipe systems. In the event of a LOCA, the reactor's emergency core cooling system (ECCS) provides core cooling to minimize fuel damage by injecting large amounts of cool, borated water into the reactor coolant system from a storage tank. The borated water stops the fission process by absorbing neutrons, and thus aids in shutting down the reactor.

The ECCS on the PWR consist of four separate systems: the high-pressure injection (or charging) system, the intermediate pressure injection system, the cold leg accumulators, and the low-pressure injection system (residual heat removal). The high pressure injection system provides water to the core during emergencies in which reactor coolant system pressure remains relatively high (such as small breaks in the reactor coolant system, steam break accidents, and leaks of reactor coolant through a steam generator tube to the secondary side). The intermediate pressure injection system is designed to accommodate emergency conditions under which the primary pressure stays relatively high; for example, small to intermediate size primary breaks. The cold leg accumulators operate without electrical power by using a pressurized nitrogen gas bubble on the top of tanks that contain large amounts of borated water. The low-pressure injection system removes residual heat by injecting water from the refueling water storage tank into the reactor coolant system during large breaks (which would cause very low reactor coolant-system pressure).

Containment Structure Designs

All U.S. reactors are surrounded by a primary containment structure that is designed to minimize releases of radioactive material into the environment. The PWR primary containment structure must surround all the components of the primary cooling system, including the reactor vessel, steam generators, and pressurizer. BWR primary containments typically are smaller, because there are no steam generators or pressurizers.

Containments must be strong enough to withstand the pressure created by large amounts of steam that may be released from the reactor cooling system during an accident. The largest containments are designed to provide sufficient space for steam released by an accident to expand and cool to keep pressure within the design parameters of the structure. Smaller containments, such as those for BWRs, require pressure suppression systems to condense much of the released steam into water. Smaller PWR containments also may include pressure suppression systems, such as ice condensers.[20]

Table 3. Containment Building Design Parameters.

Containment Type, plant	Parameter	Technical Specification
SP-1, Zion	Containment capability pressure Upper bound spike pressure Early failure physically unreasonable best estimate pressure rise, including heat sinks Time to failure, best estimate with unlimited water in cavity	149 psia[a] 107 psia 10 psi/hour 16 hours
SP-2, Surry	Containment capability pressure Upper bound spike pressure Time to failure, early failure physically unreasonable best estimate with dry cavity	134 psia 107 psia Several days
SP-3, Sequoyah	Containment capability pressure Upper bound loading pressure Lower bound loading pressure Thermal loads Early failure	65 psia, 330 °F 70-100 psia 50-70 psia 500-700 °F Quite likely
SP-4, Browns Ferry	Containment capability pressure Upper bound loading pressure Lower bound loading pressure Thermal loads Early failure	132 psia, 330 °F 132 psia in 40 minutes 132 psia in 2 hours 500-700 °F Quite likely
SP-6, Grand Gulf	Containment capability pressure Upper bound loading pressure Wall heat flux Penetration seal temperature Pressurization failure from diffusion flames Seal failure	75 psia 30 psia 1,000 to 10,000 Btu/hr-square foot 345 °F Unreasonable Unlikely
SP-15, Limerick	Containment capability pressure Upper bound loading pressure Lower bound loading pressure Thermal loads Early failure	155 psia, 330 °F 145 psia in 2-3 hours 100 psia in 3 hours 500-700 °F Rather unlikely

Source: U.S. NRC, *General Studies of Nuclear Reactors; BWR Type Reactors; Containment; Reactor Accidents; Leaks; PWR Type Reactors; Accidents; Reactors; Water Cooled Reactors; Water Moderated Reactors, NUREG-1037, 1985*, as cited by M. Ragheb UICU.

Notes: NUREG-1037 was never released, but draft versions were apparently circulated.

a. psia = pounds per square inch atmospheric.

To further limit the leakage from the containment structure following an accident, a steel liner that covers the inside surface of the containment building acts as a vapor-proof membrane to prevent any gas from escaping through any cracks that may develop in the concrete of the containment structure. Two systems act to reduce temperature and pressure within the containment structure: a fan cooler system that circulates air through heat exchangers, and a containment spray system.

All U.S. PWR designs include a containment system with Multiple Engineered Safety Features (ESFs).[21] A dry containment system consists of a steel shell surrounded by a concrete biological shield that protects the reactor against outside elements, for example, debris driven by hurricane winds or an aircraft strike.[22] The outer shield is not designed as a barrier against the release of radiation. Although the concrete structures in existing plants act as insulators against uncontrolled releases of radioactivity to the environment, they will fail if the ESFs fail in their function. Some containment building design features are summarized in **Table 3**.

The NRC Containment Performance Working Group studied containment buildings in 1985 to estimate their potential leak rates as a function of increasing internal pressure and temperature associated with severe accident sequences involving significant core damage.[23] It indentified potential leak paths through containment penetration assemblies (such as equipment hatches, airlocks, purge and vent valves, and electrical penetrations) and their contributions to leakage from for the containment. Because the group lacked reliable experimental data on the leakage behavior of containment penetrations and isolation barriers at pressures beyond their design conditions, it relied on an analytical approach to estimate the leakage behavior of components found in specific reference plants that approximately characterize the various containment types.

Nuclear Power Plants Operating in the United States

The locations of all 104 nuclear power plants operating in the United States are shown on the map in Figure 6.

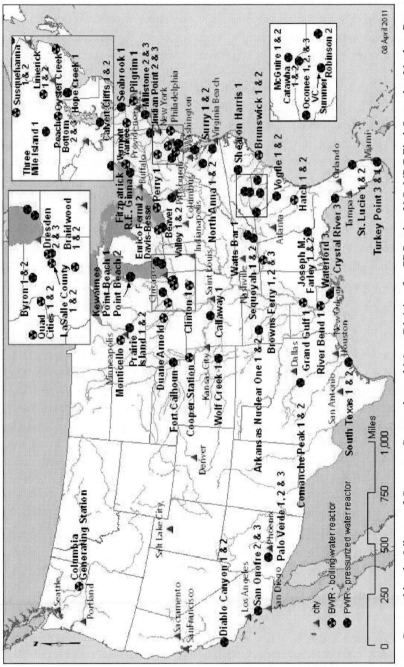

Source: Prepared by the Library of Congress Geography and Maps Division for CRS using U.S. NRC Find Operating Nuclear Reactors by Location or Name, http://www.nrc.gov/info-finder/reactor/index.html#AlphabeticalList.

Notes:

Unit	Type	MW	Vendor	St.	Lic.	Unit	Type	MW	Vendor	St.	Lic.	Unit	Type	MW	Vendor	St.	Lic.
Arkansas Nuclear 1	PWR	843	B&W	AK	1974	Grand Gulf 1	BWR	1,297	GET6	MS	1984	Point Beach 1	PWR	512	W2L	WI	1970
Arkansas Nuclear 2	PWR	995	CE	AK	1974	Hatch 1	BWR	876	GET4	GA	1974	Point Beach 2	PWR	514	W2L	WI	1973
Beaver Valley 1	PWR	892	W3L	PA	1976	Hatch 2	BWR	883	GET4	GA	1978	Prairie Island 1	PWR	551	W2L	MN	874
Beaver Valley 2	PWR	846	W3L	PA	1987	Robinson 2	PWR	710	W3L	SC	1970	Prairie Island 2	PWR	545	W2L	MN	1974
Braidwood 1	PWR	1,178	W4L	IL	1987	Hope Creek 1	BWR	1,061	GET4	NJ	1986	Quad Cities 1	BWR	867	GET3	IL	1972
Braidwood 2	PWR	1,152	W4L	IL	1988	Indian Point 2	PWR	1,023	W4L	NY	1973	Quad Cities 2	BWR	869	GET3	IL	1972
Browns Ferry 1	BWR	1,065	GET4	AL	1973	Indian Point 3	PWR	1,025	W4L	NY	1975	R. E. Ginna	PWR	498	W2L	NY	1969
Browns Ferry 2	BWR	1,104	GET4	AL	1974	Joseph M. Farley 1	PWR	851	W3L	AL	1977	River Bend 1	BWR	989	GET6	LA	1985
Browns Ferry 3	BWR	1,115	GET4	AL	1976	Joseph M. Farley 2	PWR	860	W3L	AL	1981	Salem 1	PWR	1,174	W4L	NJ	1976
Brunswick 1	BWR	938	GET4	NC	1976	Kewaunee	PWR	556	W2L	WI	1973	Salem 2	PWR	1,130	W4I	NJ	1981
Brunswick 2	BWR	937	GET4	NC	1974	LaSalle County 1	BWR	1,118	GET5	IL	1982	San Onofre 2	PWR	1,070	CE	CA	1982
Byron 1	PWR	1,164	W4L	IL	1985	LaSalle County 2	BWR	1,120	GET5	IL	1983	San Onofre 3	PWR	1,080	CE	CA	1983
Byron 2	PWR	1,136	W4L	IL	1987	Limerick 1	BWR	1,134	GET4	PA	1985	Seabrook 1	PWR	1,295	W4L	NH	1990
Callaway 1	PWR	1,236	WFL	MO	1984	Limerick 2	BWR	1,134	GET4	PA	1989	Sequoyah 1	PWR	1,148	W4L	TN	1980
Calvert Cliffs 1	PWR	873	CE	MD	1974	McGuire 1	PWR	1,100	W4L	NC	1981	Sequoyah 2	PWR	1,126	W4L	TN	1981
Calvert Cliffs 2	PWR	862	CE	MD	1976	McGuire 2	PWR	1,100	W4L	NC	1983	Shearon Harris 1	PWR	900	W3L	NC	1986

Unit	Type	MW	Vendor	St.	Lic.	Unit	Type	MW	Vendor	St.	Lic.	Unit	Type	MW	Vendor	St.	Lic.
Catawba 1	PWR	1,129	W4L	SC	1985	Millstone 2	PWR	884	CE	CT	1975	South Texas 1	PWR	1,410	W4L	TX	1988
Catawba 2	PWR	1,129	W4L	SC	1986	Millstone 3	PWR	1,227	W4L	CT	1986	South Texas 2	PWR	1,410	W4L	TX	1989
Clinton 1	BWR	1,065	GET6	IL	1987	Monticello	BWR	579	GET3	MN	1970	St. Lucie 1	PWR	839	CE	FL	1976
Columbia Gen. St.	BWR	1,190	GET5	WA	1984	Nine Mile Pt. 1	BWR	621	GET2	NY	1974	St. Lucie 2	PWR	839	CE	FL	1983
Comanche Peak 1	PWR	1,200	W4L	TX	1990	Nine Mile Pt. 2	BWR	1,140	GET5	NY	1987	Surry 1	PWR	799	W3L	VA	1972
Comanche Peak 2	PWR	1,150	W4L	TX	1993	North Anna 1	PWR	981	W3L	VA	1978	Surry 2	PWR	799	W31	VA	1973
Cooper Station	BWR	830	GET4	NE	1974	North Anna 2	PWR	973	W3L	VA	1980	Susquehanna 1	BWR	1,149	GET4	PA	1982
Crystal River 3	PWR	838	B&WLL	FL	1976	Oconee 1	PWR	846	B&WLL	SC	1973	Susquehanna 2	BWR	1,140	GET4	PA	1984
Davis-Besse	PWR	893	B&WLL	OH	1977	Oconee 2	PWR	846	B&WLL	SC	1973	Three Mile Isl. 1	PWR	786	B&WLL	PA	1974
Diablo Canyon 1	PWR	1,151	W4L	CA	1984	Oconee 3	PWR	846	B&WLL	SC	1974	Turkey Point 3	PWR	720	W3L	FL	1972
Diablo Canyon 2	PWR	1149	W4L	CA	1985	Oyster Creek	BWR	619	GET2	NJ	1991	Turkey Point 4	PWR	720	W31	FL	1973
Donald C. Cook 1	PWR	1,009	W4L	MI	1974	Palisades	PWR	778	CE	MI	1971	VC Summer	PWR	966	W31	SC	1982
Donald C. Cook 2	PWR	1,060	W4L	MI	1977	Palo Verde 1	PWR	1,335	CES80	AZ	1985	Vermont Yankee	BWR	510	GET4	VT	1972
Dresden 2	BWR	867	GET3	IL	1991	Palo Verde 2	PWR	1,335	CES80	AZ	1986	Vogtle 1	PWR	1,109	W4L	GA	1987
Dresden 3	BWR	867	GET3	IL	1971	Palo Verde 3	PWR	1,335	CES80	AZ	1987	Vogtle 2	PWR	1,127	W4L	GA	1989

(Continued)

Unit	Type	MW	Vendor	St.	Lic.	Unit	Type	MW	Vendor	St.	Lic.	Unit	Type	MW	Vendor	St.	Lic.
Duane Arnold	BWR	640	GET4	IA	1974	Peach Bottom 2	BWR	1,112	GET4	PA	1973	Waterford 3	PWR	1,250	CE	LA	1985
Fermi 2	BWR	1,122	GET4	MI	1985	Peach Bottom 3	BWR	1,112	GET4	PA	1974	Watts Bar 1	PWR	1,123	W4l	TN	1996
Fitzpatrick	BWR	852	GET4	NY	1974	Perry 1	BWR	1,261	GET6	OH	1986	Wolf Creek 1	PWR	1,166	W4L	KS	1985
Fort Calhoun	PWR	500	CE	NE	1973	Pilgrim 1	BWR	685	GET3	MA	1972						

Notes: No commercial nuclear power plants operate in Alaska or Hawaii. B&W: Babcock & Wilcox 2-Loop Lower; CE: Combustion Engineering; CE80: Combustion Engineering System 80; W2L: Westinghouse 2-Loop; W3L Westinghouse 3-Loop; W4L Westinghouse 4-Loop; GET2: General Electric Type 2; GET3: General Electric Type 3; GET4: General Electric Type 4; GET5: General Electric Type 5; GET6: General Electric Type 6.

Figure 6. Commercial Nuclear Power Plants Operating in the United States One hundred and four (104) Operating Reactors.

PLANT SEISMIC SITING CRITERIA

Earthquakes occur when stresses in the earth exceed the strength of a rock mass, creating a fault or mobilizing an existing fault.[24] The fault can slip laterally (a strike/slip fault, such as the San Andreas Fault), move vertically (a thrust or reverse fault, such as the fault that caused the March 11 Japanese earthquake), or move in some combination of the two. The fault's sudden release sends seismic shock waves through the earth that have two primary characteristics: amplitude—a measure of the peak wave height, and period—the time interval between the arrival of successive peaks or valleys.[25] The seismic wave's arrival causes ground motion. The ground motion intensity depends on three factors: the distance from the source (also known as focus or epicenter), the amount of energy released (magnitude of the earthquake), and the type of soil or rock at the site.

The shallower the earthquake's focus, the stronger the waves will be when they reach the surface. Generally, the intensity of ground shaking diminishes with increasing distance from the earthquake focus. The earthquake's magnitude (M) is measured on a logarithmic scale (sometimes referred to as the Richter scale), thus an M 7.0 earthquake has amplitude that is ten times larger than an M 6.0, but releases 31.5 times more energy than an M 6.0 earthquake. Sites with deep, soft soils or loosely compacted fill will experience stronger ground motion than sites with stiff soils, soft rock, or hard rock.

Refer to Appendix A of this report for additional discussion on magnitude, Richter scale, and intensity.

General Design Criteria

For nuclear power plants granted construction permits during the 1960s and 1970s, a design approach emerged for considering seismic loads based on site-specific investigations of local and regional seismology, geology and geotechnical engineering.[26] The 1973 publication of 10 C.F.R. 100, *Appendix A—Seismic and Geologic Siting Criteria for Nuclear Power Plants,* included the concept of a "safe shutdown earthquake" (SSE), which is discussed in a later section of this report.

General design criteria for nuclear power plants require that structures and components important to safety be designed to withstand the effects of earthquakes, tornados, hurricanes, floods, tsunamis, and seiche[27] waves without losing the capability to perform their safety function. These "safety-related" structures, systems, and components are those necessary to assure:

1. the integrity of the reactor coolant pressure boundary,
2. the capability to shut down the reactor and maintain it in a safe condition, or
3. the capability to prevent or mitigate the consequences of accidents, which could result in potential offsite exposures.

Refer to this report's section on "Nuclear Power Plant Designs" for some discussion of safety-related components.

The language in 10 C.F.R. 100, *Appendix A,* notes that the seismic criteria are based on limited geophysical and geologic information, available at the time, on faults and earthquake occurrences, and that the information would be revised when more information became available. The information is based on a review of historical records and a site investigation. Ultimately, the investigation provides the basis for determining a "safe shutdown earthquake," alternately referred to as the "design basis earthquake," defined as the maximum vibratory ground motion for which certain structures, systems, and components are designed to remain functional. Under an "operating basis earthquake," the reactor could continue operation without undue risk to the safety of the public.

The NRC subsequently published a series of Regulatory Guides in support of *Appendix A* of 10 C.F.R. 100. These guides provide technical information, procedures, and design criteria that are beyond the scope of this report.

- Regulatory Guide 1.60, *Design Response Spectra of Nuclear Power Reactors* (1973), provides ground design response spectral shapes for horizontal and vertical ground movements developed from a statistical analysis of response spectra of past Western United States (WUS) strong-motion earthquakes collected from a variety of different site conditions, primarily at deep soil sites.
- Regulatory Guide 1.165, *Identification and Characterization of Seismic Sources and Determination of Safe Shutdown Earthquake Ground Motion* (1997) ,provided procedures for (1) conducting geological, geophysical, seismological, and geotechnical investigations, (2) identifying and characterizing seismic sources, (3) conducting probabilistic seismic hazard analysis (PSHA), and (4) determining the safe shutdown earthquake for satisfying the

requirements of 10 C.F.R. 100.23. The guide evolved out of investigations into seismic hazard estimates for nuclear power plant sites operating in the Central and Eastern United States (CEUS).

- NUREG/CR-6926, *Evaluation of the Seismic Design Criteria in ASCE/SEI Standard 43-05 for Application to Nuclear Power Plants* (2007), provided seismic design criteria for safety-related structures, systems, and components in a broad spectrum of nuclear facilities.[28]

Site Investigations

The site investigations required under 10 C.F.R. 100, *Appendix A*, starts with a review of pertinent literature and progresses to field investigations. The required investigations include:

- Vibratory Ground Motion—examines lithology, stratigraphy, structural geology, underlying tectonic structures, physical earthquake evidence, engineering properties of underlying soil and rock, historically reported earthquakes, earthquake epicenters within 200 miles of site, faults within 200 miles.
- Surface Faulting—evaluates lithology, stratigraphy, structural geology, underlying tectonic structures, evidence of fault offsets, nearby faults greater than 1,000 feet in length, records of earthquakes associated with faults greater than 1,000 feet in length, epicenters of earthquakes with faults greater than 1,000 feet in length.
- Seismically Induced Floods and Water Waves — looks at reports or evidence of distantly and locally generated waves or tsunamis which have or could have affected the site, and evidence for seismically induced floods and water waves that have or could have affected the site.

Safe Shutdown Earthquake Condition

The NRC defines the Safe Shutdown Earthquake as the maximum earthquake potential for which certain structures, systems, and components, important to safety, are designed to sustain and remain functional.[29] During an earthquake, ground motion sets up vibrations in a nuclear power plant's foundation and structure. In simple terms, the vibrations represent the back-

and-forth acceleration of an object (the distance moved is the amplitude). Vibration, or horizontal ground acceleration, is measured in terms of the earth's gravitational acceleration constant (g) for structural design purposes.[30] These vibrations place additional loads and displacements on the nuclear power plant's structure, equipment and piping systems. The additional loading must be accounted for in the structural design of the piping systems supports.

Various plant structures, depending upon their elevation above the foundation, vibrate at different frequencies during an earthquake. Low frequency vibrations in the range of 1 to 10 Hz (cycles per second) are particularly problematic for a wide range of structures because such structures are often susceptible to damaging resonance at those frequencies. These accelerations and the corresponding shaking frequencies are used in the probabilistic seismic hazard analysis (PSHA) discussed in this report's "Background on Seismic Standards" section. The full seismic spectrum can be characterized by two intervals: peak ground acceleration (PGA) and spectral acceleration (SA) averaged between 5 and 10 Hz. PGA has been widely used to develop nuclear power plant "fragility estimates" and represents the performance of nuclear plant structures, systems, and components (SSCs) that are sensitive to inertial effects.

The maximum vibratory accelerations of the Safe Shutdown Earthquake must take into account the characteristics of the underlying soil material in transmitting the earthquake-induced motions at the various locations of the plant's foundation. A multiple degree-of-freedom analysis is used to simulate the effect of the earthquake on the piping systems.

Experimental and empirical seismic data have provided insights into the behavior of different structures under various acceleration and shaking conditions. One conclusion reached regarded the performance of welded steel piping at power plants during strong motion earthquakes. Relatively small numbers of failures occurred when peak ground accelerations remained below 0.5g.[31] Other types of structures would exhibit different behaviors, and engineers design the various plant structures to withstand a certain severity of earthquake specific to each plant site.

The example of **Figure 7** shows areas susceptible to shaking of a frequency of 5 Hz having a 5% probability of occurring at least once within 50 years.[32] The map shows the strength of the expected acceleration (in g) for areas experiencing such an earthquake. The darker colors on the map indicate areas of strongest shaking.

0.2-s SA with 5% in 50 year PE. BC rock. 2008 USGS

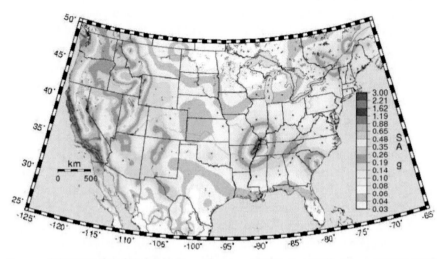

Source: USGS National Seismic Hazard maps, USGS Open-File Report 2008-1128, 2008, http://earthquake.usgs.gov/hazards/.

Notes: Areas that are susceptible to shaking at a frequency of 5 Hz with a 5% probability of occurring at least once within 50 years. The strength of the expected acceleration is expressed in terms of earth's gravitational acceleration constant (g) for areas experiencing such an earthquake. The darker colors on the map indicate areas of strongest shaking.

Figure 7. Spectral Acceleration 5 Hz Return Period of 5% in 50 Years.

National Seismic Hazard Maps

In 2008, the U.S. Geological Survey (USGS) released an update of the National Seismic Hazard Maps (NSHM).[33] The purpose of the maps is to show the likelihood of a particular severity of shaking within a specified time-period. The Seismic Hazard maps are the basis for seismic design provisions of building codes to allow buildings, highways, and critical infrastructure to withstand earthquake shaking without collapse. The NRC requires that every nuclear plant be designed for site-specific ground motions that are appropriate for their site locations. In addition, the NRC has specified a minimum ground motion level to which nuclear plants must be designed. (See discussion above on design criteria.)

The USGS revises the NHSM every six years to reflect newly published earthquake data to update building code seismic design provisions. USGS notes that the 2008 hazard maps differ significantly from the 2002 maps in many parts of the United States:

> The new maps generally show 10- to 15-percent reductions in acceleration across much of the Central and Eastern United States [CEUS] for 0.2-s [second] and 1.0-s spectral acceleration and peak horizontal ground acceleration for 2-percent probability of exceedance in 50 years. The new maps for the Western United States [WUS] indicate about 10-percent reductions for 0.2-s spectral acceleration and peak horizontal ground acceleration and up to 30-percent reductions in 1.0-s spectral acceleration at similar hazard levels.[34]

In the Central and Eastern United States (CEUS), the New Madrid Seismic Zone and the Charleston area in southeast South Carolina comprise the dominant seismic hazard (at 2% probability of exceedance in 50 years). Seismically active portions of eastern Tennessee and some portions of the northeast also contribute to the seismic hazard. The hazard at the 2% probability of exceedance in 50 years level is typically a factor of two to four times higher than the 10% probability of exceedance in 50 years values in the seismically active portions of the CEUS.

Seismic hazards are greatest in the Western United States (WUS), particularly in California, Oregon, and Washington, as well as Alaska and Hawaii. The hazard at the 2% probability of exceedance in 50 years level is typically a factor of 1.5 to 2 times higher than the 10% in 50 years values in coastal California and from 2 to 3.5 higher across the rest of the WUS.

CRS has mapped the proximity of plant sites to seismic hazards based on the USGS National Seismic Hazard Map for the United States in Figure 8. This map displays quantitative information about seismic ground motion hazards as horizontal ground acceleration (g) of a particle at ground level moving horizontally during an earthquake.

CRS has also mapped the proximity of plant sites to Quaternary period faults based on the USGS Quaternary Fault and Fold Database of the United States in Figure 9. The USGS Database has information on faults and associated folds in the United States that are believed to be sources of greater than magnitude 6 earthquakes during the past 1,600,000 years — the Quaternary period of the geologic time scale. The map is not a prediction of an earthquake event.

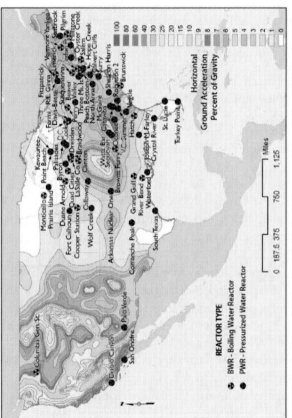

Source: Background map USGS Seismic Hazard Map for the United States, prepared for CRS by the Library of Congress Geography and Maps Division.

Notes: This map displays quantitative information about seismic ground motion hazards as horizontal ground acceleration (in terms of gravitational acceleration) of a particle at ground level moving horizontally during an earthquake. This map is not a prediction of an earthquake event. The NRC does not rank nuclear plants by seismic risk. No commercial nuclear power plants operate in either Alaska or Hawaii.

Figure 8. Operating Nuclear Power Plant Sites and Seismic Hazard Seismic hazard expressed as horizontal ground acceleration (shown as a percent of gravity).

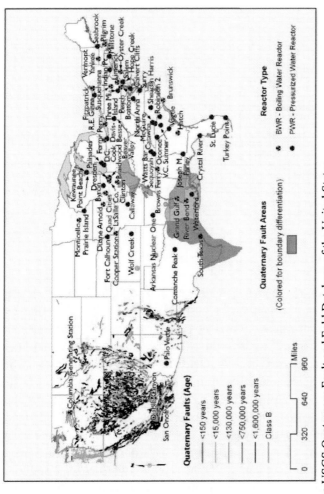

Source: CRS and the USGS Quaternary Fault and Fold Database of the United States.

Notes: To map the proximity of plant sites to faults, CRS referred to the USGS Quaternary Fault and Fold Database of the United States. This is information on faults and associated folds in the United States that are believed to be sources of greater than moment magnitude 6 (M>6) earthquakes during the Quaternary (the past 1,600,000 years). This map is not a prediction of an earthquake event. No commercial nuclear power plants operate in either Alaska or Hawaii.

Figure 9. Operating Nuclear Power Plant Sites and Mapped Quaternary Faults.

**Table 4. Operating Nuclear Power Plants Subject to Earthquake
Safety Reviews.**

Plant	St.	Type	Plant	St.	Type	Plant	St.	Type
Crystal River 3	FL	PWR	North Anna 1 & 2	VA	PWR	Sequoyah 1 & 2	TN	PWR
Dresden 2 & 3	IL	BWR	Oconee 1, 2 & 3	SC	PWR	Seabrook	NH	PWR
Duane Arnold	IA	BWR	Perry 1	OH	BWR	V.C. Summer	SC	PWR
Joseph M. Farley 1 & 2	AL	PWR	Peach Bottom 2 & 3	PA	BWR	Watts Bar 1	TN	PWR
Indian Point 2 & 3	NY	PWR	River Bend 1	LA	BWR	Wolf Creek	KS	PWR
Limerick 1 & 2	PA	BWR	Saint Lucie 1 & 2	FL	PWR			

Source: The Energy Daily.

Note: The NRC has not announced a schedule for completing the seismic reviews at the time of this report.

NRC Priority Earthquake Safety Review

The NRC has required that each nuclear plant be built to certain structural specifications based on the earthquake susceptibility of each plant site, but some of those design specifications may be reevaluated in light of new seismic analysis in the United States. In 2010 the NRC published GI-199 Safety/Risk Assessment, a two-stage assessment that determines the implications of USGS updated probabilistic seismic hazards in the Central and Eastern U.S. (CEUS) on existing nuclear power plant sites.[35] The assessment first evaluated the change in seismic hazard with respect to previous estimates at individual NPPs, and then estimated the change in Seismic Core Damage Frequency (SCDF) resulting from change in the seismic hazard. Seismic core damage frequency is the probability of damage to the reactor core (fuel rods) resulting from a seismic initiating event. It does not imply either a core meltdown or the loss of containment, which would be required for radiological release to occur. The seismic hazard at each plant site depends on the unique seismology and geology surrounding the site. Consequently, the report separately determined the implications of updated probabilistic seismic hazard for each of the 96 operating NPPs in the CEUS.[36]

The NRC does not rank nuclear plants by seismic risk. NRC's objective in the GI-199 Safety/Risk Assessment was to evaluate the need for further investigations of seismic safety for operating reactors in the CEUS. The data evaluated in the assessment suggest that the probability for earthquake ground motion above the seismic design basis for some nuclear plants in the CEUS, although still low, is larger than previous estimates. In late March 2011, the NRC announced that it had identified 27 nuclear reactors operating in the CEUS that would receive priority earthquake safety reviews.[37] Those 27 reactors are listed in Table 4.

RECENT LEGISLATIVE ACTIVITIES

Within a few days following Japan's nuclear crisis, Democrats on the House Energy and Commerce Committee requested a hearing on U.S. Nuclear Power Plant Safety and Preparedness.[38]

On March 17, 2011, the Senate Committee on Homeland Security and Governmental Affairs held a hearing on Catastrophic Preparedness that looked at technologies and emergency procedures used in the event of a large-scale earthquake or other natural disaster.[39] On April 6, 2011, the Subcommittee on Oversight and Investigations of the House Energy and Commerce Committee held a hearing the U.S. Government Response to the Nuclear Power Plant Incident in Japan.[40] On April 7, 2011, the Subcommittee on Technology and Innovation of the House Science, Space, and Technology Committee held a hearing on Earthquake Risk Reduction.[41]

Several bills have been introduced in the 112th Congress that are relevant to either nuclear power plant safety of earthquake hazard assessment.

S. 646, the Natural Hazards Risk Reduction Act of 2011, would amend the Earthquake Hazards Reduction Act of 1977 (42 U.S.C. 7704) to add program activities to research and develop effective methods, tools, and technologies to reduce the risk posed by earthquakes, and authorize the United States Geological Survey to conduct research and other activities necessary to characterize and identify earthquake hazards, assess earthquake risks, monitor seismic activity, and provide real-time earthquake information.

H.R. 1379, the Natural Hazards Risk Reduction Act of 2011, would also amend the Earthquake Hazards Reduction Act of 1977 (42 U.S.C. 7704) to research and develop effective methods, tools, and technologies to reduce the risk posed by earthquakes to the built environment, especially to lessen the risk to existing structures and lifelines.

H.R. 1268, the Nuclear Power Licensing Reform Act of 2011, would amend Section 103 of the Atomic Energy Act of 1954 (42 U.S.C. 2133), subsection c, by adding at the end the following: 'Any such renewal shall be subject to the same criteria and requirements that would be applicable for an original application for initial construction, and the Commission shall ensure that any changes in the size or distribution of the surrounding population, or seismic or other scientific data not available at time of original licensing, have not resulted in the facility being located at a site at which a new facility would not be allowed to be built.

H.R. 1242, the Nuclear Power Safety Act of 2011, would amend the Atomic Energy Act to revise regulations to ensure that nuclear facilities licensed under the act can withstand and adequately respond to an earthquake, tsunami (for a facility located in a coastal area), strong storm, or other event that threatens a major impact to the facility; a loss of the primary operating power source for at least 14 days; and a loss of the primary backup operating power source for at least 72 hours.

APPENDIX A. MAGNITUDE, INTENSITY, AND SEISMIC SPECTRUM

Earthquake magnitude is a measure of the strength of the earthquake as determined from seismographic observations. Magnitude is essentially an objective, quantitative measure of the size of an earthquake.[42] The magnitude can be expressed in various ways based on seismographic records (e.g., Richter Local Magnitude, Surface Wave Magnitude, Body Wave Magnitude, and Moment Magnitude). Currently, the most commonly used magnitude measurement is the Moment Magnitude (M) which is based on the strength of the rock that ruptured, the area of the fault that ruptured, and the average amount of slip.[43] Moment is a physical quantity proportional to the slip on the fault times the area of the fault surface that slips; it is related to the total energy released in the earthquake. The moment can be estimated from seismograms (and from geodetic measurements). The Moment Magnitude provides an estimate of earthquake size that is valid over the complete range of magnitudes, a characteristic that was lacking in other magnitude scales, such as the Richter scale.

Because of the logarithmic basis of the scale, each whole number increase in magnitude represents a tenfold increase in measured amplitude; as an estimate of energy, each whole number step in the magnitude scale corresponds to the release of about 31 times more energy than the amount associated with the preceding whole number value.

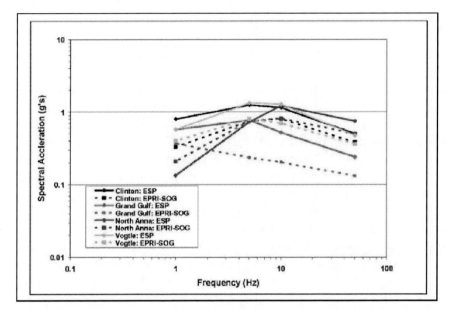

Source: NRC Generic Issue -99, *Implications of Updated Probabilistic Seismic Hazard Estimates in Central and Eastern United States on Existing Plants*, Figure 1, August 2010.

Notes: For illustrative purposes only. Originally prepared to compare seismic hazard results for four early site permit submittals. Solid line represents submittals to 1989. Dashed Lines represent Electric Power Research Institute Seismicity Owners Group Study.

Figure A-1. Spectral Acceleration (*g*) vs. Frequency (Hz) Curves are response spectral values (5% damping) at an annual exceedance frequency of 10^{-5}.

The Richter magnitude scale was developed in 1935 by Charles F. Richter of the California Institute of Technology and was based on the behavior of a specific seismograph that was manufactured at that time. The instruments are no longer in use and therefore the Richter magnitude scale is no longer used in the technical community. However, the Richter Scale is a term that is so commonly used by the public that scientists generally just answer questions about "Richter" magnitude by substituting moment magnitude without correcting the misunderstanding.

The intensity of an earthquake is a qualitative assessment of effects of the earthquake at a particular location. The intensity assigned is based on observed effects on humans, on human-built structures, and on the earth's surface at a particular location. The most commonly used scale in the United States is the Modified Mercalli Intensity (MMI) scale, which has values ranging from I to XII in the order of severity. MMI of I indicates an earthquake that was not felt except by a very few, whereas MMI of XII indicates total damage of all works of construction, either partially or completely. While an earthquake has only one magnitude, intensity depends on the effects at each particular location.

Greater magnitude earthquakes are generally associated with greater lengths of fault ruptures.[44] A fault break of 100 miles might be associated with an M8 earthquake, while a break of several miles might generate an M6 earthquake. The length of the fault break, however, is not directly proportional to the energy released. The induced amplitude of acceleration (g) does increase with increasing magnitude (M). Various methods have been developed to relate the magnitude of an earthquake to the amplitude of acceleration it induces, and different methods may result in significant variations in results.

The seismic spectrum can be characterized by two intervals—peak ground acceleration (PGA) and spectral acceleration averaged between 5 and 10 Hz (SAAvg5-10). PGA has been widely used to develop fragility estimates and represents the performance of nuclear plant structures, systems, and components (SSCs) that are sensitive to inertial effects.

Figure A-1 shows a example of response spectra for several power plants.[45] The frequency range, of 1 to 10 *Hz*, is the subject of USGS earthquake hazard studies, as discussed above.

NUREG/CR-6926 references the American Society of Civil Engineers (ASCE) Standard 7-05 *Minimum Design Loads for Buildings and Other Structures* seismic hazard maps. The maximum considered earthquake (MCE) is based on spectral accelerations with 2%/50 yr probability (2% probability of being equaled or exceeded in any single year in 50 years or otherwise stated as a 2% annual exceedance probability). (To obtain the design earthquake spectral response accelerations (DS) used in structural design, the spectral accelerations are multiplied by 2/3.) At sites in seismically active regions in the Western United States (WUS), the corresponding DS hazard is approximately 10%/50 yr (return period of 475 yr). In the Central and Eastern United States (CEUS) this hazard is approximately 4%/50 yr (return period of approximately 1,200 yr), These are due to differences in the typical slopes of seismic hazard curves in the WUS and CEUS.

APPENDIX B. TERMS

Boiling water reactor (BWR) directly generates steam inside the reactor vessel.

Deterministic Seismic Hazard Assessment (DSHA) focuses on a single earthquake event to determine the finite probability of occurring.

Double-ended guillotine break (DEGB) represents a break of the largest diameter pipe in the primary system that the emergency core cooling system (ECCS) must be sized to provide adequate makeup water to compensate for.

Light water reactor systems use ordinary water as a fuel moderator and coolant, and uranium fuel artificially enriched to 4.5%-5% fissile uranium-235. Includes BWR and PWR types.

Loss of Coolant Accident (LOCA) is the most severe operating condition for a reactor that can contribute to a reactor core meltdown.

Operating Basis Earthquake is the maximum vibratory ground motion that a reactor could continue operation without undue risk and safety of the public.

Pressurized water reactor (PWR) uses two major loops to convert the heat generated by the reactor core into steam outside of the reactor vessel.

Probabilistic Seismic Hazard Assessments (PSHA) attempt to quantify the probability of exceeding various ground-motion levels at a site given all possible earthquakes.

Safe Shutdown Earthquake (also design basis earthquake) is the maximum vibratory ground motion at which certain structures, systems, and components are designed to remain functional.

Seismic Core Damage Frequency is the probability of damage to the core resulting from a seismic initiating event.

ACKNOWLEDGMENTS

Jacqueline C. Nolan, Library of Congress, Geography and Maps Division

Richard J. Campbell, Specialist in Energy Policy, Congressional Research Service

Peter Folger, Specialist in Energy and Natural Resources Policy, Congressional Research Service

End Notes

[1] This report does not discuss the risk from earthquake-caused tsunamis, as associated with the catastrophic damage to the Fukushima plants.

[2] Heavy water reactors, such as Canada's CANDU reactor, use water enriched with a heavier hydrogen isotope and natural uranium for fuel, which contains less than 3.5% uranium-235.

[3] For further background uranium fuel, see CRS Report RL34234, *Managing the Nuclear Fuel Cycle: Policy Implications of Expanding Global Access to Nuclear Power*, coordinated by Mary Beth Nikitin.

[4] Office of Technology Assessment, *Nuclear Power Plant Standardization: Light Water Reactors*, NTIS order #PB81-213589, April 1981, p. 11.

[5] Julian J. Bommer, Norman A. Abrahamson, and Fleur O. Strasser, et al., "The Challenge of Defining Upper Bounds on Earthquake Ground Motions," *Seismological Research Letters*, vol. 75, no. 1 (February 2004).

[6] R. J. Budnitz, G. Apostolakis, and D. M. Boore, *Recommendations for Probabilistic Seismic Hazard Analysis: Guidance on Uncertainty and Use of Experts: Main Report*, U.S. Nuclear Regulatory Commission, Nureg/CR-6372, Lawrence Berkeley National Laboratory, CA, April 1997, http://www.nrc.gov/reading-rm/doc-collections/nuregs/contract/cr6372/vol1/index.html#pub-info.

[7] Edward (Ned) H. Field, *Probabilistic Seismic Hazard Analysis (PSHA) - A Primer*, http://www.relm.org/ tutorial_materials.

[8] U.S. Nuclear Regulatory Commission, *Implications of Updated Probabilistic Seismic Hazard Estimates in Central and Eastern United States Existing Plants - Safety/Risk Assessment*, Generic Issue 199 (GI-199), August 2010.

[9] U.S. NRC, NRC frequently asked questions related to the March 11, 2011 Japanese Earthquake and Tsunami, March 2011, http://www.nrc.gov.

[10] U.S. Nuclear Regulatory Commission, *Reactor Concepts Manual, Boiling Water Reactor Systems*, http://www.nrc.gov/reading-rm/basic-ref/teachers/03.pdf - 2005-10-17.

[11] During the sustained chain reaction in an operating reactor, the U-235 splits into highly radioactive fission products, while the U-238 is partially converted to plutonium-239 by neutron capture, some of which also fissions. Further neutron capture creates other radioactive elements. The process of radioactive decay transforms an atom to a more stable element through the release of radiation—alpha particles (two protons and two neutrons), charged beta particles (positive or negative electrons), or gamma rays (electromagnetic radiation).

[12] The NRC regulates the design, construction, and operation requirements of the ECCS under 10 C.F.R. 50.46, "Acceptance criteria for emergency core cooling systems for light-water nuclear reactors"; Appendix K to 10 C.F.R. Part 50, "ECCS Evaluation Models"; and Appendix A to 10 C.F.R. Part 50, "General Design Criteria [GDC] for Nuclear Power Plants" (e.g., GDC 35, "Emergency Core Cooling").

[13] N.C. Chokshi, S.K. Shaukat, and A.L. Hiser, et al., *Seismic Considerations for the Transition Break Size*, U.S. Nuclear Regulatory Commission, NUREG 1903, Brookhaven National Laboratory, February 2008.

[14] U.S. NRC, *Reactor Concepts Manual, Pressurized Water Reactor Systems,* http://www.nrc.gov/reading-rm/basic-ref/teachers/04.pdf - 2005-10-17.

[15] The two-loop units in the United States are Ginna, Kewaunee, Point Beach 1 and 2, and Prairie Island 1 and 2.

[16] The three-loop units in the United States are Beaver Valley 1 and 2, Farley 1 and 2, H. B. Robinson 2, North Anna 1 and 2, Shearon Harris 1, V. C. Summer, Surry 1 and 2, and Turkey Point 3 and 4.

[17] The four-loop units in the United States are Braidwood 1 and 2, Byron 1 and 2, Callaway, Catawba 1 and 2, Comanche Peak 1 and 2, D. C. Cook 1 and 2, Diablo Canyon 1 and 2, Indian Point 2 and 3, McGuire 1 and 2, Millstone 3, Salem 1 and 2, Seabrook, Sequoyah 1 and 2, South Texas Project 1 and 2, Vogtle 1 and 2, Watts Bar 1, and Wolf Creek.

[18] The Babcock & Wilcox units in the United States are Arkansas 1, Crystal River 3, Davis Besse, Oconee 1, 2, and 3, and Three Mile Island 1.

[19] The Combustion Engineering units in the United States are Arkansas 2, Calvert Cliffs 1 and 2, Fort Calhoun, Millstone 2, Palisades, Palo Verde 1, 2, and 3, San Onofre 2 and 3, Saint Lucie 1 and 2, and Waterford 3.

[20] Kazys Almenas and R. Lee, *Nuclear Engineering: An Introduction* (Berlin: Springer-Verlag, 1992), pp. 507-514.

[21] M. Ragheb, *Containment Structures* (2011). University of Illinois Champaign-Urbana, https://netfiles.uiuc.edu/mragheb/www/NPRE%20457%20CSE%20462%20 Safety% 20Analysis%20of%20Nuclear%20Reactor%20Systems/ Containment%20Structures.pdf.

[22] NRC regulations require that new reactors be designed to withstand the impact of large commercial aircraft and that existing plants develop strategies to mitigate the effects of large aircraft crashes. See CRS Report RL34331, *Nuclear Power Plant Security and Vulnerabilities*, by Mark Holt and Anthony Andrews.

[23] U.S. Nuclear Regulatory Commission, *General Studies of Nuclear Reactors; BWR Type Reactors; Containment; Reactor Accidents; Leaks; PWR Type Reactors; Accidents; Reactors; Water Cooled Reactors; Water Moderated Reactors*, NUREG-1037, May 1, 1985.

[24] The Applied Technology Council (ATC) and the Structural Engineers Association of California (SEAOC), *Briefing Paper 1 Building Safety and Earthquakes Part A: Earthquake Shaking and Building Response*, Redwood City, CA, http://www.atcouncil.org/.

[25] The wave's frequency is the inverse of the period $(1/s)$, and is expressed as the number of wave cycles per second (termed Hertz or *Hz*).

[26] U.S. Nuclear Regulatory Commission, *Evaluation of the Seismic Design Criteria in ASCE/SEI Standard 43-05 for Application to Nuclear Power Plants*, NUREG/CR-6926, Brookhaven National Laboratory, NY, March 2007.

[27] Standing waves, or waves that move vertically but not horizontally. Seiche waves can be triggered by earthquakes, strong winds, tides, and other causes.

[28] Based on a review by the American Society of Civil Engineers/Structural Engineering Institute (ASCE/SEI) Standard 43-05 - *Seismic Design Criteria for Structures, Systems, and Components in Nuclear Facilities*.

[29] http://www.nrc.gov/reading-rm/basic-ref/glossary/safe-shutdown-earthquake.html

[30] Gravitation acceleration g = 32 feet/second/second (ft/second^2).

[31] N.C. Chokshi, S.K. Shaukat, and A.L. Hiser, et al., *Seismic Considerations for the Transition Break Size*, U.S. NRC, NUREG-1903, February 2008, pp. 29-30.

[32] This collection of USGS seismic hazard maps includes probabilistic ground motion maps for Peak Ground Acceleration (PGA), 1Hz (1.0 second SA), and 5Hz (0.2 second SA). (Refer to the report section on "Safe Shutdown Earthquake" for a discussion of spectral acceleration.) Some additional spectral accelerations (SA) are also included for central and

southern California. Most figures correspond to the 2% in 50-year probability of exceedance, but there are a few figures for the 10% in 50 year and the 5% in 50-year probability of exceedance as well a range of accelerations and associated probabilities.

[33] Mark D. Petersen, Arthur D. Frankel, and Stephen C. Harmsen, et al., *Documentation for the 2008 Update of the United States National Seismic Hazard Maps*, U.S Geological Survey, Open-File Report 2008-1128, 2008, http://earthquake.usgs.gov/hazards/.

[34] Ibid.

[35] U.S. Nuclear Regulatory Commission, *Implications of Updated Probabilistic Seismic Hazard Estimates in Central and Eastern United States Existing Plants—Safety/Risk Assessment*, Generic Issue 199 (GI-199), August 2010.

[36] Ibid.

[37] George Lobsenz, "NRC Task Force To Review Safety: 27 Reactors Are Seismic Priorities," *The Energy Daily*, March 24, 2011.

[38] House Committee on Energy & Commerce Democrats, Committee Democrats Request Hearing on U.S. Nuclear Power Plant Safety and Preparedness, http://democrats.energycommerce.house.gov/index.php?q=news/committee-democrats-request-hearing-on-us-nuclear-power-plant-safety-and-preparedness.

[39] Senate Committee on Homeland Security & Governmental Affairs, Catastrophic Preparedness: How Ready is FEMA for the Next Big Disaster? http://hsgac.senate.gov/public/index.cfm?FuseAction=Hearings.Hearing&Hearing_ID= a42880b1-22fc-4890-b82c-dd2a369e2aa2

[40] House Energy & Commerce Committee, The U.S. Government Response to the Nuclear Power Plant Incident in Japan, http://energycommerce.house.gov/hearings/hearingdetail.aspx?NewsID=8420.

[41] House Committee on Science, Space, and Technology, Subcommittee Reviews Status of U.S. Earthquake Preparedness, http://science.house.gov/press-release/subcommittee-reviews-status-us-earthquake-preparedness.

[42] US NRC, *NRC frequently asked questions related to the March 11, 2011 Japanese Earthquake and Tsunami.*

[43] USGS, *Measuring Earthquakes*, http://earthquake.usgs.gov/learn/faq/?categoryID=2&faqID=23.

[44] H. Bolton Seed, I. M. Idriss, and Fred. W. Kiefer, "Characteristics of Rock Motions During Earthquakes," *Journal of Soil Mechanics and Foundation Division, Proceedings of the American Society of Civil Engineers*, September 1969, pp. 1199-1217.

[45] Frequency *Hz* (Hertz) refers to the number of cycles per second (which is inverse of the ground motion wave period — the time between two wave peaks). Thus, 0.2-*s* is the equivalent of 5 *Hz* (1/0.2-*s*), and 1-*s* is the equivalent of 1 Hz (1/1-*s*).

In: Nuclear Power Plants   ISBN: 978-1-61470-952-7
Editor: James P. Argyriou   ©2012 Nova Science Publishers, Inc.

Chapter 2

SEISMIC ISSUES FOR NUCLEAR POWER PLANTS

U.S. Nuclear Regulatory Commission

Nuclear power plants are built to withstand environmental hazards, including earthquakes. Even those plants that are located outside of areas with extensive seismic activity are designed for safety in the event of such a natural disaster. The Nuclear Regulatory Commission (NRC) requires all of its licensees to take seismic activity into account when designing and maintaining its nuclear power plants. When new seismic hazard information becomes available, the NRC evaluates the new data and models and determines if any changes are needed at plants. The newest seismic data suggests that although the potential seismic hazard at some nuclear power plants in central and eastern states may have increased beyond previous estimates, all operating nuclear plants remain safe with no need for immediate action.

BACKGROUND

The agency requires plant designs to withstand the effects of natural phenomena including earthquakes (i.e., seismic events). The agency's requirements, including General Design Criteria for licensing a plant, are described in Title 10 of the *Code of Federal Regulations* (10 CFR). These license requirements include traditional engineering practices such as "safety margins." Practices such as these add an extra element of safety into design, construction, and operations.

The NRC has always required licensees to design, operate, and maintain safety-significant structures, systems, and components to withstand the effects of earthquakes and to maintain the capability to perform their intended safety functions. The agency ensures these requirements are satisfied through the licensing, reactor oversight, and enforcement processes.

EARTHQUAKE (OR SEISMIC) HAZARD

The NRC requires that safety-significant structures, systems, and components be designed to take into account:

- The most severe natural phenomena historically reported for the site and surrounding area. The NRC then adds a margin for error to account for the limited historical data accuracy;
- Appropriate combinations of the effects of normal and accident conditions with the effects of the natural phenomena; and
- The importance of the safety functions to be performed.

The U.S. Geological Survey (USGS) Web site provides general information about earthquakes (*http://earthquake.usgs.gov/learning*. An earthquake releases energy that radiates from the fault and causes ground movement. As the ground moves, objects such as nuclear power plant structures on or in the ground also move. The nature of the movement depends on how the earthquake releases energy and location. The intensity of an earthquake can be characterized by both the frequency of the shaking and by the acceleration of the ground at the plant. These characteristics describe how the energy released from the earthquake impacts the plant's buildings as well as the systems and components that are housed and supported by those buildings.

Earthquake characteristics provide information used in designing existing nuclear plants. The frequency of the shaking is measured in cycles per second (or Hz), and the acceleration is typically expressed as some fraction of the acceleration of gravity, which is about 32.2 feet per second per second (ft/s^2). For example, an acceleration of 0.15 g (15 percent of the acceleration of gravity) is about equal to an acceleration of 5 ft/s^2.

SEISMIC SAFETY ASSESSMENT

The licensing bases for existing nuclear power plants considered historical data at each site. This data is used to determine design basis loads from the area's maximum credible earthquake, with an additional margin included. In Generic Letter 88-20, the NRC required existing plants to assess their potential vulnerability to earthquake events, including those that might exceed the design basis, as part of the Individual Plant Examination of External Events Program. This process was intended to examine the available safety margins of existing plants beyond the design basis (Safe Shutdown Earthquake) and to report on certain modifications of identified seismic vulnerabilities.

Today, the NRC utilizes a risk-informed regulatory approach, including insights from probabilistic assessments and traditional deterministic engineering methods to make regulatory decisions about existing plants (e.g., licensing amendment decisions). Any new nuclear plant the NRC licenses will use a probabilistic, performance-based approach to establish the plant's seismic hazard and the seismic loads for the plant's design basis.

ADDITIONAL MEASURES FOLLOWING SEPT. 11, 2001

Following the events of September 11, 2001, NRC required all nuclear plant licensees to take additional steps to protect public health and safety in the event of a large fire or explosion. If needed, these additional steps could also be used during natural phenomena such as earthquakes, tornadoes, floods, and tsunami. In general, these additional steps are plans, procedures, and pre-staged equipment whose intent is to minimize the effects of adverse events. In accordance with NRC regulations, all nuclear power plants are required to maintain or restore cooling for the reactor core, containment building, and spent fuel pool under the circumstances associated with a large fire or explosion. These requirements include using existing or readily available equipment and personnel, having strategies for firefighting, operations to minimize fuel damage, and actions to minimize radiological release to the environment.

EVOLVING KNOWLEDGE ABOUT EARTHQUAKES

The central and eastern United States (CEUS) is generally an area of low to moderate earthquake hazard with few active faults in contrast to the western United States. Even so, in 181 1–1812, three major earthquakes (Magnitude 7 to 7.7 on the commonly used Richter scale) shook much of the CEUS. These earthquakes occurred near the town of New Madrid, M.O. In 1886, a large earthquake (Richter scale magnitude of about 7) occurred near Charleston, S.C. This earthquake caused extensive damage and was felt in most of the eastern United States. Geologists are aware of these historic occurrences, and knowledge of such earthquakes was taken into account in plant design and analysis.

The NRC regularly reviews new information on earthquake source and ground motion models. For example, the NRC reviewed updated earthquake information provided by applicants in support of Early Site Permits for new reactors. This additional information included new models to estimate earthquake ground motion and updated models for earthquake sources in seismic regions such as eastern Tennessee and around both Charleston and New Madrid.

The NRC examined 2008 earthquake-related information to assess potential safety implications for nuclear power plants in central and eastern states. Analysis of these updates indicated slight increases to earthquake hazard estimates for some plants in the CEUS. The NRC also reviewed and evaluated recent USGS earthquake hazard estimates for the CEUS that are used for building code applications outside of plant licensing. These reviews showed that the seismic hazard estimates at some current CEUS operating sites may potentially be slightly higher than what was expected during design and previous evaluations, although there is adequate protection at all plants.

NRC RESPONSE TO INCREASED ESTIMATED CEUS EARTHQUAKE HAZARDS

The NRC began assessing the safety implications of increased plant earthquake hazards in 2005 when the staff recommended examining the new CEUS earthquake hazard information under the Generic Issues Program (GIP). The NRC staff identified the issue as GI-199 and completed a limited scope screening analysis for the seismic issue in December 2007, to decide whether additional review is needed. The screening compared the new seismic data with earlier seismic evaluations conducted by the NRC staff. This analysis confirmed that operating nuclear power plants remain safe with no need for immediate action. The assessment also found that, although overall seismic risk remains low, some seismic hazard estimates have increased and warrant further attention. In September 2010, NRC issued a Safety/Risk Assessment report and an Information Notice (http://www.nrc.gov/reading- to inform stakeholders of the assessment results.

The NRC is developing a Generic Letter (GL) to request information from all U.S. nuclear plants. The GL will be issued in draft form to support a public meeting in late spring 2011. NRC expects to issue the GL by the end of 2011, near the time when new seismic models will become available. These new seismic models are being developed by NRC, the U.S. Department of Energy, and the Electric Power Research Institute and will be reviewed by the USGS. The NRC expects to receive information from the GL in 2012 and will review it to determine whether any plant improvements are needed.

Information regarding this generic issue and the GIP in general is available at http://www.nrc.gov/about-nrc/regulatory/gen-issues

INSPECTIONS FOLLOWING JAPAN EVENT

The NRC is not currently performing inspections that are directly related to GI-199. However, on March 23, 2011, the NRC directed its inspectors to assess the actions taken by nuclear plant licensees in response to events at the Fukushima Daiichi nuclear station in Japan. NRC inspectors will perform inspections to verify that important equipment and materials are adequate and properly staged, tested, and maintained in order to respond to a severe earthquake, flooding event, or loss of all electrical power. Inspections are expected to be completed by the end of April 2011. The results will be issued in an inspection report by May 13 that will be made publicly available.

To read more about risk-related NRC policy, see the Probabilistic Risk Assessment Fact Sheet (http://www.nrc.gov/reading and Nuclear Reactor Risk (http://www.nrc.gov/reading-rm/doc-collections/fact-sheets/reactor-risk.html). Each provides more information on the use of probability in evaluating hazards (including earthquakes) and their potential impact on plant safety margins. Questions and answers on the March 2011 earthquake and tsunami are available at http://www.nrc.gov/japan/faqs-related-to-japan.pdf.

In: Nuclear Power Plants
Editor: James P. Argyriou

ISBN: 978-1-61470-952-7
©2012 Nova Science Publishers, Inc.

Chapter 3

NEW NUCLEAR PLANT DESIGNS

U.S. Nuclear Regulatory Commission

BACKGROUND

The NRC has long sought standardization of nuclear power plant designs, and the enhanced safety and licensing reform that standardization could make possible. The Commission expects advanced reactors to be safer and use simplified, passive or other innovative means to accomplish their safety functions. The NRC's regulation (Part 52 to Title 10 of the Code of Federal Regulations) provides a predictable licensing process including certification of new nuclear plant designs. This process reflects decades of experience and research involving reactor design and operation. The design certification process provides for early public participation and resolution of safety issues prior to an application to construct a nuclear power plant.

PRE-APPLICATION REVIEW PROCESS

The NRC's "Statement of Policy for Regulation of Advanced Nuclear Power Plants," dated July 8, 1986, encourages early discussions, before a license application is submitted, between NRC and reactor designers to provide licensing guidance. In June 1988, the NRC issued NUREG-1226, "Development and Utilization of the NRC Policy Statement on the Regulation of Advanced Nuclear Power Plants." This document provides guidance on the implementation of the policy and describes the approach used by NRC in its review of advanced reactor design concepts.

In general, the NRC conducts pre-application reviews of advanced reactor designs to indentify:

- major safety issues that could require Commission policy guidance to the staff,
- major technical issues that could be resolved under existing NRC regulations on policy, and
- research needed to resolve identified issues.

DESIGN CERTIFICATION REVIEW PROCESS

The review process for new reactor designs involves certifying standard reactor designs, independent of a specific site, through a rulemaking (Subpart B of Part 52). This rulemaking can certify a reactor design for 15 years. Design certification applicants must provide the technical information necessary to demonstrate compliance with the safety standards set forth in applicable NRC regulations (10 CFR Parts 20, 50, 73, and 100). Applicants must also provide information to close out unresolved and generic safety issues, as well as issues that arose after the Three Mile Island accident. The application must include a detailed analysis of the design's vulnerability to certain accidents or events, and inspections, tests, analyses, and acceptance criteria to verify the key design features. The NRC is considering a new rule that would require design certification applicants to assess their plant's level of built-in protection for avoiding or mitigating the effects of a large commercial aircraft impact, reducing the need for human intervention to protect public.

Currently there are four certified reactor designs that can be referenced in an application for a combined license (COL) to build and operate a nuclear power plant. They are:

1. Advanced Boiling Water Reactor design by GE Nuclear Energy (May 1997);
2. System 80+ design by Westinghouse (formerly ABB-Combustion Engineering) (May 1997);
3. AP600 design by Westinghouse (December 1999); and
4. AP1000 design (pictured at left) by Westinghouse (January 2006).

REACTOR DESIGN REVIEW STATUS

The status of advanced reactor applications for both active and inactive design reviews is provided below in alphabetical order. A description of each design follows.

Active Reviews

- *AP1000* (Amendment) – Westinghouse submitted an application to amend the AP1000 design in July 2007, in order to: 1) address several "open items" that would otherwise be dealt with in a COL application, 2) voluntarily comply with the intent of the proposed aircraft impact assessment rule, and 3) modify the reactor's pressurizer design. The staff accepted the amendment for review in January 2008 and expects to complete its review in 2009. The rulemaking is tentatively scheduled for completion in 2010.
- *ESBWR* - General Electric submitted its ESBWR (pictured at right) certification application on Aug. 24, 2005. The staff accepted the application for review in a letter dated Dec. 1, 2005, and expects the certification process to continue through 2010.

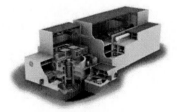

- *EPR* - Areva submitted its EPR certification application on Dec. 11, 2007. The staff expects the certification process to continue through 2011.
- *US-APWR* - Mitsubishi Heavy Industries (MHI), a Japanese firm, submitted a design certification application for the U.S.-specific version of its Advanced Pressurized Water Reactor (pictured at right) on Dec. 31, 2007. The staff expects the certification process to continue through 2011.

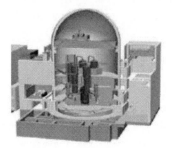

Pre-Application Reviews

- *PBMR* - A South African firm, Pebble Bed Modular Reactor (PBMR) Pty. Limited notified the NRC in a letter dated Feb. 18, 2004, that it intended to apply for design certification in the near future and requested discussions with the NRC to plan the scope and content of the preapplication review. NRC staff have held several public meetings with PBMR to discuss its activities and plans to submit pre-application information. PBMR has continued to submit pre-application information through 2007 and expects to submit a design certification application in late 2009.
- *Toshiba 4S* - On Feb. 2, 2005, the NRC staff met with the city manager and vice mayor of Galena, Alaska to discuss and answer questions on the city's plans to build a Toshiba 4S reactor to provide its electricity. Toshiba began pre-application discussions with NRC staff in Oct. 2007, and the company expects to submit a design approval application in 2009.

Inactive Reviews

- *IRIS* - In May 2006, Westinghouse and the NRC staff discussed the current status of the International Reactor Innovative and Secure (IRIS). The planned submittal of a design certification application for IRIS has been changed from 2008 to 2010. Westinghouse has submitted topical reports related to the planned test programs and plans to submit additional reports in support of preapplication interactions. The IRIS design is sometimes mentioned in the context of a grid- appropriate reactor under the Global Nuclear Energy Partnership.

REGULATORY STRUCTURE FOR NEW PLANT LICENSING

In the longer term, the NRC may be called on to review reactor designs that use a broader range of technology than those currently under review. Therefore, the NRC staff may develop technology-neutral guidelines for plant licensing in the future. These guidelines are intended to encourage future designs to incorporate additional safety and security where possible. The staff issued in Dec. 2007 a "Feasibility Study for a Risk-Informed and Performance-Based Regulatory Structure for Future Plant Licensing" *(NUREG-1 860).*

Design Descriptions

- *ABWR:* The U.S. Advanced Boiling Water Reactor (pictured at left) uses a single-cycle, forced circulation design with a rated power of 1,300 megawatts electric (MWe). The design incorporates features of the BWR designs in Europe, Japan, and the United States, and uses improved electronics, computer, turbine, and fuel technology. Improvements include the use of internal recirculation pumps, control rod drives that can be controlled by a screw mechanism rather than a step process, microprocessor-based digital control and logic systems, and digital safety systems. The design also includes safety enhancements such as protection against overpressurizing the containment, passive core debris flooding capability, an independent water makeup system, three emergency diesels, and a combustion turbine as an alternate power source.

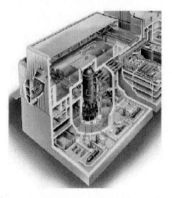

- *AP600:* The Advanced Passive 600 is a 600 MWe advanced pressurized water reactor that incorporates passive safety systems and simplified system designs. The passive systems use natural driving forces without active pumps, diesels, and other support systems after actuation. Use of redundant, non-safetyrelated, active equipment and systems minimizes unnecessary use of safety-related systems.
- *AP1000:* The Advanced Passive 1000 is a larger version of the previously approved AP600 design. This 1,100 MWe advanced pressurized water reactor incorporates passive safety systems and simplified system designs. It is similar to the AP600 design but uses a longer reactor vessel to accommodate longer fuel, and also includes larger steam generators and a larger pressurizer.
- *EPR:* The Evolutionary Power Reactor (pictured at right) is a 1,600 MWe pressurized water reactor of evolutionary design. Design features include four 100% capacity trains of engineered safety features, a double-walled containment, and a "core catcher" for containment and cooling of core materials for severe accidents resulting in reactor vessel failure. The design does not rely on passive safety features. The first EPR is under construction at the Olkiluoto site in Finland, with another planned for the Flammanville site in France.
- *ESBWR:* The Economic and Simplified Boiling Water Reactor is a 1,500 MWe, natural circulation boiling water reactor that incorporates passive safety features. This design is based on its predecessor, the 670 MWe Simplified BWR (SBWR) and also utilizes features of the

certified ABWR. The ESBWR enhances natural circulation by using a taller vessel, a shorter core, and by reducing the flow restrictions. The design utilizes an isolation condenser system for high-pressure water level control and decay heat removal during isolated conditions. After the automatic depressurization system operates, a gravity-driven cooling system provides low-pressure water level control. Containment cooling is provided by a passive system.

- *IRIS:* The International Reactor Innovative and Secure is a pressurized light water cooled, medium-power (335 MWe) reactor that has been under development for several years by an international consortium. The IRIS design utilizes an integral reactor coolant system layout. The IRIS reactor vessel houses not only the nuclear fuel, control rods and neutron reflector, but also all the major reactor coolant system components including pumps, steam generators and pressurizer. The IRIS integral vessel is larger than a traditional PWR pressure vessel, but the size of the IRIS containment is a fraction of the size of corresponding loop reactors.
- *PBMR:* The Pebble Bed Modular Reactor (pictured at left) is a modular high-temperature gas reactor that uses helium as its coolant. PBMR design consists of eight reactor modules, 165 MWe per module, with capacity to store 10 years of spent fuel in the plant (there is additional storage capability in onsite concrete silos). The PBMR core is based on German high-temperature gas-cooled reactor technology and uses spherical graphite elements containing ceramic-coated fuel particles.

- *System 80+:* This standard plant design uses a 1,300 MWe pressurized water reactor. It is based upon evolutionary improvements to the standard CE System 80 nuclear steam supply system and a balance-ofplant design developed by Duke Power Co. The System 80+ design has safety systems that provide emergency core cooling, feedwater and decay heat removal. The new design also has a safety depressurization system for the reactor, a combustion turbine as an alternate AC power source, and an in- containment refueling water storage tank to enhance the safety and reliability of the reactor system.
- *Toshiba 4S:* The Toshiba 4S reactor design has an output of about 10 MWe. The reactor has a compact core design, with steel-clad metal-alloy fuel. The core design does not require refueling over the 30-year lifetime of the plant. A three-loop configuration is used: primary system (sodium-cooled), an intermediate sodium loop between the radioactive primary system and the steam generators, and the water loop used to generate steam for the turbine. The basic layout is a "pool" configuration, with the pumps and intermediate heat exchanger inside the primary vessel.
- *US-APWR:* The Mitsubishi Heavy Industry US-APWR design is an evolutionary 1,700 MWe pressurized water reactor currently being licensed and built in Japan. The design includes high-performance steam generators, a neutron reflector around the core to increase fuel economy, redundant core cooling systems and refueling water storage inside the containment building, and fully digital instrumentation and control systems.

In: Nuclear Power Plants
Editor: James P. Argyriou

ISBN: 978-1-61470-952-7
©2012 Nova Science Publishers, Inc.

Chapter 4

THE COMMERCIAL OUTLOOK FOR U.S. SMALL MODULAR NUCLEAR REACTORS

U.S. Department of Commerce

In May 2009, President Barack Obama called for harnessing the power of nuclear energy "on behalf of our efforts to combat climate change and to advance peace and opportunity for all people." Meeting the energy, environment, and climate demands of the 21st century will require creating new solutions and reimagining older but still crucial technologies. Civil nuclear technology combines elements of both approaches.

Although large reactors that produce in excess of 1,000 megawatts of electricity (MWe) are the most common, they are not the only possible designs for power stations. In fact, small modular reactors (SMRs), many of which are built on proven, well-known technology, could meet the power and heating needs of countries worldwide while significantly reducing carbon emissions. SMRs could also help U.S. companies increase exports and create new jobs. And some SMRs could be designed to safeguard nuclear technology and fuel from falling into hostile hands.

This report provides an overview of U.S. SMR development, examines the strengths of the technology, looks at the market characteristics that correlate best with SMRs, and identifies in greater detail the obstacles to SMR deployment. The report will conclude with recommendations for U.S. policy-makers and industry to facilitate the safe and secure deployment of SMRs in the medium to long term and to contribute to national and energy security, climate change mitigation, and economic growth.

SMR designs are promising to industry as a commercial opportunity and to governments as a way to meet growing energy needs with non–fossil fuel–based energy. They must be deployed, however, with the same safety and security measures as for larger reactors. Because of the long life of nuclear reactors, the nature of nuclear waste, and the serious consequences of any potential accident, a thorough design certification and licensing process must be in place. While decades of experience and a world-class regulatory system have given Americans comfort that nuclear power can be safely and securely deployed, evaluating the new U.S. SMR designs will take time and cooperation between vendors and the Nuclear Regulatory Commission (NRC).

Some significant challenges to eventual SMR deployment exist. Some of those barriers relate to foreign markets, such as the need for additional bilateral nuclear cooperation agreements with foreign countries, intense foreign competition (often from state-owned enterprises), and the lack of a global nuclear liability regime. Other obstacles are domestic. Those issues include the erosion of U.S. nuclear manufacturing capacity and the need for strong government assistance, such as manufacturing tax credits and loan guarantees specifically for manufacturers. Although technical hurdles remain before SMRs will be ready for commercial use, overcoming the other obstacles will be critical to the eventual deployment of U.S. SMRs.

SMR OVERVIEW

The International Atomic Energy Agency (IAEA) defines small reactors as producing equivalent electric power less than 300 MWe, although larger reactors could also be categorized as modular (that is, able to be assembled from standardized, mass-produced subcomponents). In the SMR context, *modular* essentially means that the main components of the reactor are constructed in an off-site factory and are delivered to the plant under construction for final assembly and installation. In contrast to small reactors, most reactors deployed in the United States have electric power capacities between 800–1,200 MWe. Most of the newer large reactor designs, such as the Westinghouse AP1000 and the GE-Hitachi ESBWR, have capacities between 1,100–1,600 MWe. Traditionally, those reactors have been built from the ground up on site; however, an increasing amount of modular components are being used in the newest large reactors.

In many ways, the history of small reactors began in the late 1940s and 1950s when the United States pursued using nuclear reactors to power military operations and naval vessels. The *U.S.S. Nautilus*, the first nuclear-powered submarine, was launched in 1955, and today the U.S. Navy operates 82 vessels powered by 103 nuclear reactors.[1] A one-of-a-kind nuclear-powered civilian ship, the *NS Savannah*, was launched in 1962, also using a small reactor to power the vessel. It was used for goodwill tours and even cargo runs before being taken out of service in 1970.[2] Other countries, including China, France, Russia, and the United Kingdom, also operate nuclear naval programs. The U.S. Army also experimented with using small reactors to power military bases and to provide electricity in remote areas; however, the program was halted in the mid-1970s.

Naval reactors are typically pressurized water reactors (PWRs), produce 50–300 MWe depending on ship type, and can be refueled far less frequently than existing civilian reactors. (They are allowed by the Department of Defense to use more highly enriched fuel than civilian reactors.)

In some ways, current U.S. SMR proposals resemble the naval reactors that have been operating for decades.

A number of U.S. SMR designs are currently being pursued. They share a number of essential characteristics that set them apart from existing reactor designs: (a) smaller dimensions and scaled-down electricity-generating capacity, (b) reduced number of components, and (c) modularity and potential for factory assembly and rail-shipping. Some SMR designs also include new fuel cycle approaches and new safety and security postures (including passive safety features[3]).

SMR designs can also be generally categorized into two groups: those based on existing light water reactor designs, and those that use a coolant other than water. The second group can be further divided into gas-cooled reactors and liquid metal– cooled reactors.

TERMINOLOGY NOTE: ACTIVE AND PASSIVE SAFETY FEATURES

Most existing reactors have active safety features, which are electromechanical devices that are engineered to provide reactor cooling and to shut off other systems in the event of a reactor incident. Passive safety features, however, rely on physical forces or properties, such as temperature or gravity, rather than the operation of specific equipment to prevent serious accidents. These features reduce the risk of failure inherent in engineered components.[4]

Light-water reactors (LWRs) use ordinary water as the coolant. The most common types of LWRs are boiling water reactors and PWRs.[5] U.S. SMR designs that also use ordinary water as a coolant include the Babcock & Wilcox mPower, the NuScale Power Module, and the Westinghouse IRIS. The manufacturers of those designs have the advantage of decades of collective industry experience operating LWRs and have the existing manufacturing and infrastructure to support LWR production.

There is also a range of U.S. SMR designs that use coolants other than water and prepare fuel in unconventional ways. Several different government and industry consortiums (including in the United States through the Next Generation Nuclear Plant Program[6]) are pursuing high-temperature gas-cooled reactors. Those reactors use a gas, such as helium, as a coolant, rather than water. This coolant allows the reactor to operate at a higher temperature, which, in turn, produces hot gas that can (a) have its heat transferred through a heat exchanger for use in industrial applications, (b) be put through a steam generator to drive an electricity-generating turbine, or (c) be used to directly drive a turbine. Other advanced designs in the United States include liquid metal–cooled fast reactors, which use liquid metals, such as sodium or lead, as the primary coolant. GE-Hitachi's PRISM, Hyperion Power Generation's Hyperion Power Module, and Advanced Reactor Concepts' ARC- 100 are examples of liquid metal–cooled designs.

STRENGTHS OF SMRs

A primary advantage of SMRs is in their production. Their small size means that they do not need the ultra-heavy forged components that currently can be made only by Japan Steel Works and Doosan Heavy Industries in South Korea.[7] In most of the current U.S. SMR designs, the reactor pressure vessels and other large forgings could be supplied by domestic vendors, which would create U.S. jobs and potential exports of SMR components to international customers. In addition, most SMR designs allow for factory manufacturing, which could potentially provide opportunities for cost savings, for increased quality, and for more efficient production. Those attributes mean that SMRs could be a significant source of economic growth in the United States.

Some SMR advocates say that smaller reactors have a cost advantage. Conventional nuclear power plants tend to have high capital costs per MWe. Advocates say that the modular production and smaller size could lower capital costs and give quicker returns on investment. The modular nature of SMRs also means that power stations could be built in a stepwise fashion, generating electricity and revenue more quickly to pay for further expansion. Theoretically, SMRs could reduce operating, maintenance, and fuel-cycle costs, because many designs could operate for longer cycles than do existing reactors (although more frequent outages and inspections might be required for safety purposes). Additional cost savings could be realized if a smaller on-site workforce compared to that used for traditional reactors is able to provide the necessary safety and security oversight for plant operations.

In addition, some nuclear manufacturers maintain maintain that economies of scale mean large reactors are more cost-effective in the long run, even if SMR capital costs are lower. In other words, although individual SMRs might be cheaper to build than larger reactors, the larger reactors can produce significantly more electricity over long periods of time; thus, the cost per kilowatt-hour of electricity from larger reactors is lower than that for SMRs. One way this cost difference might be resolved is by building SMRs in parallel. In other words, reap the cost savings of modular manufacturing while building enough modules to duplicate the electricity output of a larger reactor.

Some SMRs could be suited for specialized applications. The small size and output of some designs could provide advantages over large nuclear units for industrial or district heating applications because using a traditional reactor would be too expensive and would produce far too much energy to be used efficiently for those purposes. SMRs could also be used for energy-intensive activities located in remote areas, such as desalination plants and certain mining operations. A similar application could be to provide heat and electricity for oil shale recovery, which is a particularly energy-intensive operation. If nuclear reactors, rather than fossil fuel–based technology, could power oil extraction from tight shale then they could significantly lower the carbon emissions from such recovery and make the extraction more attractive.

Another potential long-term strength of SMRs is that some designs could also support nuclear non-proliferation objectives. All U.S. SMRs are designed to be deployed in an underground configuration. Industry observers contend that this would limit the risk for aboveground sabotage (which is a serious consideration for traditional nuclear power plants) or for radioactive release. The fuel cycle (particularly uranium enrichment and reprocessing) is where most non-proliferation concerns lie. The U.S. SMRs likely to be deployed in the near term are similarly fueled as the existing LWRs, but some U.S. vendors argue that the United States could exercise greater influence in the global nuclear fuel trade if U.S. SMR technology were widely deployed.

TERMINOLOGY NOTE: SMRs

The abbreviation SMR is used interchangeably to mean "small or medium-sized reactor" and "small modular reactor," with the latter definition becoming increasingly dominant within the industry. Therefore, this paper defines SMR as "small modular reactor."

"SMRs can be a solution for certain markets that have smaller and less robust electricity grids and limited investment capacity."

Some U.S. SMR vendors claim that their designs could be "black boxed" (that is, they could be deployed already fueled), and once the fuel is spent, the entire unit could be shipped back to the factory for waste handling and reprocessing. If the responsibility for the fuel cycle is taken out of the hands of the reactor operator, then risks of proliferation could potentially be reduced. Significant technical issues, however, remain unsolved for this concept, and there are serious outstanding questions involving transportation, waste handling, safety, and security. Although an attractive idea, such designs are unlikely to be deployed in the near or mid term.

MARKETS RIPE FOR SMR SOLUTIONS

Traditional nuclear reactors provide base-load electricity. Those nuclear reactors are usually run on a continual and constant basis to provide electricity to meet the minimum demand, or base load, as opposed to some other power plants (such as natural gas-fired units), which are generally run to provide electricity during peak demand periods. Because nuclear power plants are typically large (both in size and electricity output), they require (a) significant upfront investment (roughly $5 billion per plant, though costs can vary widely); (b) an electricity grid capable of handling the power output of the plant; (c) the ability to responsibly manage the nuclear waste generated by the reactors; (d) a need to account for and to minimize environmental impacts; and (e) general public acceptance. The last point can be especially contentious, particularly in countries where, for historical reasons, nuclear power is viewed as more "dangerous" than other forms of energy. Those factors could significantly limit the markets in which traditional large nuclear reactors can feasibly be deployed.

The significant benefits of nuclear power (in particular the production of base-load electricity with little greenhouse gas [GHG] emissions), however, are causing countries to consider how they could deploy some type of nuclear reactor. SMRs could be a solution for certain markets that have smaller and less robust electricity grids and limited investment capacity, and thus limited ability to build the infrastructure needed for a large reactor.

SMRs might also be a good fit in markets where anticipated electricity demand is projected to increase incrementally, because SMRs could be built in series as needed. SMRs could also be used as a power solution for smaller areas that are experiencing rapid population growth and electricity demand, as a heat solution for specific industrial processes, or as an energy solution for energy-intensive activities such as desalination.

SMRs might be particularly attractive in countries that currently rely on diesel generators for producing limited amounts of expensive electricity. Small reactors could make economic sense because of the high cost of diesel generation compared to the low marginal cost of producing electricity from nuclear energy. (This advantage, however, might be eroded by the initial investment costs and the need to establish a national regulatory program.) Some SMRs could also be a solution for markets that lack the qualified engineers and skilled craft workers needed to construct large reactors on site, because they could be modularly fabricated and then delivered for assembly.

Finally, given the growing prominence of climate change concerns, one particularly attractive feature of nuclear power is its ability to produce base-load electricity with negligible GHG emissions. As developing countries in particular move toward low-GHG emissions growth strategies, SMRs could provide a low-emissions power solution. From the perspective of potential SMR vendors, countries that offer tax and other incentives for low-GHG emissions technology are particularly attractive for SMRs, if nuclear is an accepted technology under those development plans. (See the appendix for an evaluation of potential overseas markets for SMRs.)

INTERNATIONAL OBSTACLES TO U.S. SMR COMPETITIVENESS

A number of U.S. companies are pursuing SMR technology for commercial sale, including GE- Hitachi Nuclear, Westinghouse Electric Company, NuScale Power, Babcock & Wilcox, Hyperion Power Generation, Advanced Reactor Concepts, and General Atomics.

Just like exporters of traditional large reactors, U.S. SMR vendors would face intense foreign competition, primarily by state-owned or state-aligned enterprises. Foreign nuclear companies have enjoyed significant government support, ranging from direct government ownership and management to favorable financing, industrial coordination, and support for manufacturers.

Some U.S. suppliers also regard the lack of international licensing standards as an obstacle to expanding their business. They say that obtaining regulatory approval in one market does not provide any "leg up" in obtaining approval in another market, which means that the process has to be repeated for each country that the supplier wants to sell to. However, it is difficult to see how international licensing standards could be developed or enforced given the unique national circumstances that factor into a regulator's licensing decision- making. The discretion of these national regulators cannot be compromised. More generally, U.S. suppliers also say that the lack of regulatory infrastructure in many countries interested in SMR technology is a problem for ensuring the safe and secure deployment of the technology. This challenge also applies to larger, traditional reactors.

Nuclear liability is a significant concern for SMR and large reactor designers. Currently, no global nuclear liability regime exists. This situation not only complicates commercial arrangements, but also means that, in the unlikely event of a nuclear incident, claims for damages would be the subject of protracted and complicated litigation in the courts of many countries against multiple potential defendants with no guarantee of recovery. The IAEA-sponsored Convention on Supplementary Compensation for Nuclear Damage (CSC) is the only international instrument that provides the basis for establishing a global regime, including countries with and without nuclear power facilities. U.S. nuclear suppliers have stated that the implementation of CSC is a necessity for pursuing a major nuclear export program.

DOMESTIC OBSTACLES TO U.S. SMR COMPETITIVENESS

There are also domestic policies that hinder U.S. SMR competitiveness, with some policies relevant to all nuclear suppliers and some specific to SMR deployment, both at home and abroad.

One obstacle is diminished manufacturing capacity. U.S. nuclear competitiveness is hampered because U.S. manufacturing capacity has been eroded through the lack of new reactor construction during the past few decades. Some government resources to help manufacturers are not appropriate for nuclear suppliers, or the resources exclude the suppliers entirely. For example, only two U.S. nuclear manufacturers qualified for the advanced energy manufacturing tax credit. The timeline to be eligible for the credit requires a facility to be up and running four years from certification. Some U.S. firms say that the timeline is too short for many nuclear suppliers; just acquiring the high-precision machines necessary to retool and rebuild capacity can require a lead time of several years.

Some U.S. suppliers also note that the United States currently levies tariffs between 3.3 percent and 5.2 percent on key nuclear reactor components, but the tariffs are currently suspended in some cases (specifically for reactor pressure vessels and steam turbine generators that were ordered before July 31, 2006). Tariffs around the world, particularly in the European Union and South Korea, are higher on such components. Coupled with significant foreign government support, foreign suppliers can more easily enter the U.S. market, while U.S. manufacturers face a significant trade barrier in key foreign markets.

Generally, SMR vendors say that additional 123 agreements (see terminology note) are needed with new markets overseas to legally permit U.S. companies to engage in trade of major nuclear reactor components and fuel with those markets. Once the 123 agreements are in force, U.S. companies may still need to obtain authorizations and licenses from the Departments of Commerce, Energy, and State, as well as from the Nuclear Regulatory Commission (NRC). Many companies say that the process is challenging to navigate. The Department of Commerce, through its Civil Nuclear Trade Initiative, published the "Civil Nuclear Exporters Guide" in 2009 to help U.S. companies with this process.[8]

"Just like exporters of traditional large reactors, U.S. SMR vendors would face intense foreign compteition, primarily by state-owned or state-aligned enterprises."

TERMINOLOGY NOTE: 123 AGREEMENTS

The term 123 agreements refers to nuclear cooperation agreements required by section 123 of the U.S. Atomic Energy Act of 1954, as amended, before significant transfers of nuclear material, equipment, or components from the United States to another nation can take place. Without a 123 agreement in force, U.S. firms cannot engage in substantial nuclear trade with another country. Currently, the United States has bilateral agreements with 21 countries and Taiwan, plus agreements with the European Atomic Energy Community (which includes the 27 European Union member countries) and the IAEA.

According to some U.S. suppliers, several other U.S. government policies may pose challenges to SMR deployment. For example, to meet the requirements of the Omnibus Budget Reconciliation Act of 1990, as amended, the NRC assesses a uniform annual fee for each licensed nuclear power reactor under 10 CFR Part 171.[9] The total annual fee for each operating power reactor includes a spent fuel storage and reactor-decommissioning annual fee. Separate from the annual fees assessed under 10 CFR Part 171, an annual premium for the nuclear liability insurance pool is required by the Price–Anderson Act. In 2009, the NRC issued an Advance Notice of Proposed Rulemaking to consider whether to amend 10 CFR Part 171 to establish a variable annual fee structure for power reactors based on the reactor's licensed power limit contained in the operating license. If the NRC issued regulations based on a variable fee structure accounting for reactor size, then it is reasonable to assume that the annual fee assessed to SMRs would be less than the annual fee assessed to the current large LWRs.

Another consideration U.S. SMR vendors have to address is that the NRC's requirement for the emergency planning zones (EPZs) around reactors does not generally take into account the size of a reactor. SMR vendors argue that the smaller size means that a smaller protection area could suffice, which would maintain safety while providing cost savings. The NRC's regulations do allow the size of the EPZ to be adjusted on a case-by-case basis for reactors that are gas cooled (such as the high-temperature gas-cooled reactor designs mentioned earlier) or have a thermal output of less than 250 MW. This exception would cover many of the proposed SMR designs, if the vendors can demonstrate that a smaller EPZ is acceptable on the basis of their emergency planning. Adjusting the 250 MW limit could cover the rest of the U.S. SMR designs not currently eligible for this potential size exception. Aside from the size regulation, additional costs related to emergency planning stem from state and local regulations, which cover environmental protection, police and fire coverage, and other services. SMR vendors will need to work with operators and state and local authorities to determine if SMRs warrant adjustments to those other existing regulations.

Other suppliers suggest that current NRC requirements for staffing and security systems at a reactor site would be unnecessary for an SMR, because the requirements should be tied more closely to reactor size. The staffing and security requirements (colloquially referred to as "guns and guards") are a necessary expense for reactors to ensure the safe operation of the reactor and the security of the nuclear material. If the deployment of SMRs allows for reduction in those costs, SMRs could be more attractive to potential customers. For the NRC to consider adjustments to those requirements, however, SMR vendors must engage in a technical discussion with the regulator and demonstrate how the reactors could be safely and securely operated with fewer control room operators and guards. U.S. suppliers also say that they enjoy a cooperative relationship with the NRC and that progress is being made on addressing those issues.

Another obstacle is that the NRC is facing a significantly increased workload as it reviews new LWR designs and prepares to issue the first combined operating licenses for new large reactor construction in the United States. Many industry representatives say the NRC needs more funding and additional resources to properly review SMRs in a timely fashion. Yet the NRC has significantly increased its staffing during the past several years, including hiring almost 500 new staff members for the Office of New Reactors, which was established in 2006.

One additional obstacle is beyond the scope of this report but could play a significant role in whether SMRs are commercially deployed: public opinion. To the extent that the smaller profile of SMRs results in their deployment closer to population centers, public opposition to their deployment might rise. Deployment at existing sites, or in industrial applications away from residential areas, however, might minimize the impact of public opinion. Education about the safety features of SMRs and nuclear reactors in general could also ameliorate this concern.

IMPACT OF SMRS ON U.S. JOB CREATION

A serious obstacle to the resurgence of traditional nuclear power in the United States is the eroded domestic manufacturing capacity for the major nuclear components. A robust program of building SMRs, however, could make use of existing domestic capacity that is already capable of completely constructing most proposed SMR designs. SMRs would not require the ultra-heavy forgings that currently can only be made overseas. U.S. suppliers say that firms could retool using existing capabilities and resources and could source most of the components of SMRs here in the United States. This ability could mean tremendous new commercial opportunities for U.S. firms and workers.

A substantial SMR deployment program in the United States could result in the creation of many new jobs in manufacturing, engineering, trans-portation, construction (for site preparation and installation) and craft labor, professional services, and ongoing plant operations. As SMR manufacturers prove their designs in the domestic market, they will likely consider export opportunities. The modular nature of SMRs and their relative portability means that locating export-oriented SMR manufacturing and assembly could make sense for U.S. companies, as opposed to the localization that is typically necessary for building larger reactors.

OUTLOOK

Although SMRs have significant potential and the market for their deployment is growing, their designs must still go through the technical and regulatory processes necessary to ensure that they can be safely and securely deployed. Light- water technology–based SMRs may not be ready for deployment in the United States for at least a decade, and advanced designs might be even further off. Light-water SMRs and SMRs that have undergone significant testing are the most likely candidates for near-term deployment, because they are most similar to existing reactors that have certified designs and significant operating histories. NuScale is on track to submit its reactor design to the NRC by 2012, as is Babcock & Wilcox for its mPower design. In addition, GE-Hitachi, which already completed an NRC preapplication review for its PRISM reactor in 1994, plans to submit its PRISM design for certification in 2012.

With fierce competition for commercial deployment of U.S. SMRs anticipated, the U.S. government is accelerating its efforts to support the licensing of new reactor designs. The fiscal year 2011 budget request for the Department of Energy includes $39 million for a program to support design certification of SMRs for commercial deployment, as well as a research and development portfolio that will address the technology development needs of both near- and longer-term SMRs. The Department of Energy is also in discussions with several U.S. companies to facilitate the light- water SMR design certification by the NRC within a reasonable timeframe. The department also continues to support research and development efforts toward advanced reactor designs through the Advanced Reactor Concepts program, which focuses on metal-cooled reactor technologies.

As designs move closer to deployment, the Department of Commerce is ready with resources to help both U.S. SMR designers and the broader nuclear supply chain. The Department of Commerce has launched the Civil Nuclear Trade Initiative, which identifies the U.S. nuclear industry's most pressing trade policy challenges and the most significant commercial opportunities. The initiative then coordinates public- and private-sector efforts to address the opportunities and challenges in a way that supports the industry's endeavors to rebuild its manufacturing base. To accomplish the goals, the initiative includes four pillars: (a) an interagency working group on civil nuclear trade; (b) the Civil Nuclear Trade Industry Advisory Committee; (c) trade policy and promotion activities, including trade missions, official advocacy, and industry programs; and (d) stakeholder resources, such as the "Civil Nuclear Exporters Guide."[10]

"A substantial SMR deployment program in the United States could result in the creation of many new jobs."

POLICY AND INDUSTRY RECOMMENDATIONS

Policy-makers and U.S. companies can take a number of actions to move toward the commercial deployment of SMRs. For policy-makers, these include the following actions:

- Strengthen U.S. government efforts to bring the Convention on Supplementary Compensation for Nuclear Damage into force.
- Consider additional 123 agreements for markets that might be appropriate for SMRs.

- Continue to provide support to countries in their efforts to develop the regulatory infrastructure needed to ensure the safe and secure building and operation of nuclear reactors.
- Explicitly include civil nuclear projects in future clean-energy programs, such as the Advanced Energy Manufacturing Tax Credit Program, and ensure that the terms of such credits are applicable to nuclear projects (including allowing for longer lead times).
- Set aside a portion of future nuclear loan guarantee funds to support the rebuilding of U.S. nuclear manufacturing capacity.
- Support NRC's consideration of adjustments to annual assessments, EPZs, and reactor staffing and security requirements, contingent on U.S. vendors' demonstration and the NRC's evaluation that such adjustments will not compromise the safe and secure operation of nuclear reactors.

U.S. SMR companies should consider the following actions:

- Provide a list of priority markets to the U.S. government for additional 123 agreements.
- Report specific trade barriers and policy challenges, both domestic and international, to the Department of Commerce.
- Schedule preapplication reviews for SMR designs with the NRC and provide requested information in a timely manner.
- Ensure that emergency plans include detailed explanations of the technical reasons SMR designs merit NRC adjustment to some requirements, while still meeting safety and security objectives.
- Participate in U.S. government–sponsored nuclear efforts, including multilateral forums such as the International Framework for Nuclear Energy Cooperation; bilateral dialogues with key markets; trade policy and promotion activities, including trade missions and the U.S. Industry Promotion Program at the IAEA general conference; and industry advisory committees, such as the Civil Nuclear Trade Advisory Committee.

APPENDIX: POTENTIAL BEST PROSPECT MARKETS FOR SMRs

Methodology

Twenty-seven countries were identified by the Department of Commerce as markets of interest for new nuclear expansion, but the countries varied from existing nuclear powers to nations at preliminary stages of readiness to actually undertake a nuclear program. The countries were then rated on the basis of how closely they matched seven characteristics of a potential SMR market: (a) low population density, (b) anticipated population growth, (c) anticipated carbon emissions growth, (d) anticipated economic growth, (e) anticipated energy consumption growth, (f) importation of electricity, and (g) existing nuclear capacity. Countries were ranked for each category into quartiles depending on their scores (the bottom quartile has six countries instead of seven) and were assigned a score of 1–4 for the first five categories and either 1 or 0 for the latter two categories. The higher the score is, the closer the country is to the ideal characteristics of an SMR market. "Population density" is the only category in which the listed score has an inverse relationship to the underlying data, because lower population density is considered a stronger indication that SMR deployment may be appropriate.

Country	Population density[a]	CO_2 emissions[b]	Electricity imports[c]	Economic growth[d]	Energy consumption[e]	Nuclear capacity[f]	Total
Latvia	4	4	1	4	4	1	18
Turkey	2	4	0	4	4	1	15
Jordan	3	2	1	4	4	1	15
Lithuania	4	4	0	3	2	1	14
India	1	4	1	4	3	0	13
Armenia	2	3	0	4	4	0	13
China	1	4	0	4	4	0	13
United Arab Emirates	3	4	0	1	4	1	13
Morocco	3	3	0	2	4	1	13
Estonia	4	2	0	3	3	1	13
Bulgaria	3	2	0	3	3	0	11
Brazil	4	3	1	1	2	0	11
Indonesia	1	4	0	2	3	1	11
Ghana	2	2	1	3	2	1	11
South Korea	1	3	0	3	3	0	10
Nigeria	1	2	0	3	3	1	10

(Continued)

Country	Population densitya	CO2 emissionsb	Electricity importsc	Economic growthd	Energy consumpti	Nuclear capacityf	Total
Kenya	3	3	0	1	2	1	10
Mexico	3	3	0	1	2	0	9
South Africa	4	1	0	3	1	0	9
Slovak Republic	2	1	1	4	1	0	9
Ukraine	3	2	0	2	2	0	9
Poland	2	2	0	2	2	1	9
Egypt	2	1	0	2	3	1	9
Canada	4	1	0	1	1	0	7
Czech Republic	1	3	0	2	1	0	7
Slovenia	2	1	0	2	1	0	6
Netherlands	1	1	1	1	1	0	5

Explanatory notes for this table appear on page 10.

Notes: Explanatory Notes to Appendix Table

a. Calculated using 2009 Central Intelligence Agency *World Factbook* data for population and geographic area. A higher score means a lower population density. SMRs could be more appro priate for a scarcer population distribution.

b. Calculated as the percentage change in carbon dioxide (CO_2) emissions between 2004 and 2008, according to Energy Information Administration (EIA) data. A higher score means a higher percentage increase in CO_2 emissions. SMRs could serve to reduce CO_2 emissions by providing emissions-free base-load power.

c. Rated using 2008 EIA net electricity imports data. A '1" means the country is a net importer, and a '0" means the country either is a net electricity exporter or does not import or export electricity. Countries with more electricity demand than domestic generating capacity might see SMRs as an option for meeting that additional demand.

d. Calculated by averaging annual gross domestic product (GDP) growth from 2004 to 2008 using World Bank data. A higher score means higher average annual GDP growth. Countries with increasing GDP growth, particularly growth that is occurring incrementally, might use SMRs to meet concomitant increases in energy demand.

e. Calculated on the basis of growth in kilowatt-hour per capita consumption from 2004 to 2007, using World Bank data. A higher score means a higher increase in kilowatt-hour per capita consumption.

f. Rated using World Nuclear Association information. A '0" means the country has an operating reactor, and a '1" means the country does not have an operating reactor. SMRs may be an attractive option for countries with limited or no nuclear experi-ence, whereas countries with an operating reactor might find that expanding traditional-sized reactors at existing plants makes more sense.

End Notes

[1] World Nuclear Association, "Nuclear-Powered Ships," September 2010, www.world-nuclear.org/info/inf34.html.

[2] U.S. Maritime Administration, "Nuclear Ship Savannah," www.marad.dot.gov/ships_shipping_landing_page/ns_savannah_home/ns_savannah_home.htm.

[3] So-called Generation III+ large reactor designs, including the Westinghouse AP1000 and the GE-Hitachi ESBWR, also have passive safety features.

[4] World Nuclear Association, "Small Nuclear Power Reactors," October 2010, *www.world-nuclear.org/info/inf33.html*.

[5] Nuclear Regulatory Commission, "Online Glossary," August 2010, www.nrc.gov/reading.html.

[6] The Next Generation Nuclear Plant program is sponsored by a Department of Energy initiative to fund research projects in support of developing a high-temperature gas-cooled reactor.

[7] Additional capacity for ultra-heavy forging is in the planning stages in China, France, India, and the United Kingdom.

[8] The guide, which is currently being updated, is available at www.export.gov/civilnuclear.

[9] Title 10, Code of Federal Regulations, lists regulations promulgated by the NRC.

[10] For more information on the Civil Nuclear Trade Initiative, please contact ITA's civil nuclear industry specialists at civilnuclear@trade.gov.

In: Nuclear Power Plants
Editor: James P. Argyriou

ISBN: 978-1-61470-952-7
©2012 Nova Science Publishers, Inc.

Chapter 5

NUCLEAR POWER PLANT SECURITY AND VULNERABILITIES

Mark Holt and Anthony Andrews

SUMMARY

The physical security of nuclear power plants and their vulnerability to deliberate acts of terrorism was elevated to a national security concern following the attacks of September 11, 2001. Since the attacks, Congress has repeatedly focused oversight and legislative attention on nuclear power plant security requirements established and enforced by the Nuclear Regulatory Commission (NRC

The Energy Policy Act of 2005 (EPACT05, P.L. 109-58) imposed specific criteria for NRC to consider in revising the "Design Basis Threat" (DBT), which specifies the maximum severity of potential attacks that a nuclear plant's security force must be capable of repelling. In response to the legislative mandate, NRC revised the DBT (10 C.F.R. Part 73.1) on April 18, 2007. Among other changes, the revisions expanded the assumed capabilities of adversaries to operate as one or more teams and attack from multiple entry points.

To strengthen nuclear plant security inspections, EPACT05 required NRC to conduct "force-onforce" security exercises at nuclear power plants at least once every three years. In these exercises, a mock adversary force from outside a nuclear plant attempts to penetrate the plant's vital area and simulate damage to a "target set" of key safety components. From the start of the program through 2009, 112 force-on-force inspections were conducted, with each inspection typically including three mock attacks by the adversary force. During the 112 inspections, eight mock attacks resulted in the simulated destruction of complete target sets, indicating inadequate protection against the DBT, and additional security measures were promptly implemented, according to NRC.

EPACT05 also included provisions for fingerprinting and criminal background checks of security personnel, their use of firearms, and the unauthorized introduction of dangerous weapons. The designation of facilities subject to enforcement of penalties for sabotage was expanded to include waste treatment and disposal facilities.

Nuclear power plant vulnerability to deliberate aircraft crashes has been a continuing issue. After much consideration, NRC published final rules on June 12, 2009, to require all new nuclear power plants to incorporate design features that would ensure that, in the event of a crash by a large commercial aircraft, the reactor core would remain cooled or the reactor containment would remain intact, and radioactive releases would not occur from spent fuel storage pools.

NRC rejected proposals that existing reactors also be required to protect against aircraft crashes, such as by adding large external steel barriers, deciding that other mitigation measures already required by NRC for all reactors were sufficient. In 2002, NRC ordered all nuclear power plants to develop strategies to mitigate the effects of large fires and explosions that could result from aircraft crashes or other causes. NRC published a broad final rule on nuclear reactor security March 27, 2009, including fire mitigation strategies and requirements that reactors establish procedures for responding to specific aircraft threats.

Other ongoing nuclear plant security issues include the vulnerability of spent fuel pools, which hold highly radioactive nuclear fuel after its removal from the reactor, standards for nuclear plant security personnel, and nuclear plant emergency planning. NRC's March 2009 security regulations addressed some of those concerns and included a number of other security enhancements.

OVERVIEW OF REACTOR SECURITY

Physical security at nuclear power plants involves the threat of radiological sabotage—a deliberate act against a plant that could directly or indirectly endanger public health and safety through exposure to radiation. The Nuclear Regulatory Commission (NRC) establishes security requirements at U.S. commercial nuclear power plants based on its assessment of plant vulnerabilities to, and the consequences of, potential attacks. The stringency of NRC's security requirements and its enforcement program have been a significant congressional issue, especially since the September 11, 2001, terrorist attacks on the United States.

While NRC establishes security requirements within the boundaries of commercial nuclear sites, the Department of Homeland Security (DHS) has broad responsibility for coordinating government-wide efforts to prevent and respond to terrorist attacks, including attacks on nuclear power plants. DHS works with NRC and other agencies to protect nuclear facilities and other critical infrastructure.[1]

Nuclear plant security measures are designed to protect three primary areas of vulnerability: controls on the nuclear chain reaction, cooling systems that prevent hot nuclear fuel from melting even after the chain reaction has stopped, and storage facilities for highly radioactive spent nuclear fuel. U.S. plants are designed and built to prevent dispersal of radioactivity, in the event of an accident, by surrounding the reactor in a steel-reinforced concrete containment structure.

NRC requires commercial nuclear power plants to have a series of physical barriers and a trained security force, under regulations already in place prior to the 9/11 attacks (10 C.F.R. 73—Physical Protection of Plants and Materials). The plant sites are divided into three zones: an "owner-controlled" buffer region, a "protected area," and a "vital area." Access to the protected area is restricted to a portion of plant employees and monitored visitors, with stringent access barriers. The vital area is further restricted, with additional barriers and access requirements. The security force must comply with NRC requirements on pre-hiring investigations and training.[2]

A fundamental concept in NRC's physical security requirements is the design basis threat (DBT), which establishes the severity of the potential attacks that a nuclear plant's security force must be capable of repelling. The DBT includes such characteristics as the number of attackers, their training, and the weapons and tactics they could use. Specific details are classified. Critics of nuclear plant security have contended that the DBT should be strengthened to account for potentially larger and more sophisticated terrorist attacks.

Reactor vulnerability to deliberate aircraft crashes has also been a major concern since 9/11. Most existing nuclear power plants were not specifically designed to withstand crashes from large jetliners, although analyses differ as to the damage that could result. NRC has determined that commercial aircraft crashes are beyond the DBT but published regulations in June 2009 to require that new reactor designs be able to withstand such crashes without releasing radioactivity. Nuclear power critics have called for retrofits of existing reactors as well.

Since the 9/11 attacks, NRC and Congress have taken action to increase nuclear power plant security. NRC issued a series of security measures beginning in 2002, including a strengthening of the DBT and establishing the Office of Nuclear Security and Incident Response (NSIR). The office centralizes security oversight of all NRC-regulated facilities, coordinates with law enforcement and intelligence agencies, and handles emergency planning activities. In 2004, NRC implemented a program to conduct "force on force" security exercises overseen by NSIR at each nuclear power plant at least every three years. The Energy Policy Act of 2005 (P.L. 109-58) required NRC to further strengthen the DBT, codified the force-on-force program, and established a variety of additional nuclear plant security measures. In March 2009, NRC published a series of security regulations that require power plants to prepare cyber security plans, develop strategies for dealing with the effects of aircraft crashes, strengthen access controls, improve training for security personnel, and take other new security measures.

DESIGN BASIS THREAT

The design basis threat describes general characteristics of adversaries that nuclear plants and nuclear fuel cycle facilities must defend against to prevent radiological sabotage and theft of strategic special nuclear material. NRC licensees use the DBT as the basis for implementing defensive strategies at specific nuclear plant sites through security plans, safeguards contingency plans, and guard training and qualification plans.

General requirements for the DBT are prescribed in NRC regulations,[3] while specific attributes of potential attackers, such as their weapons and ammunition, are contained in classified adversary characteristics documents (ACDs).

Fundamental policies on nuclear plant security threats date back to the Cold War. In 1967, the Atomic Energy Commission (AEC) instituted a rule that nuclear plants are not required to protect against an attack directed by an "enemy of the United States."[4] That so-called "Enemy of the State Rule" specifies that nuclear power plants are

> not required to provide for design features or other measures for the specific purpose of protection against the effects of (a) attacks and destructive acts, including sabotage, directed against the facility by an enemy of the United States, whether a foreign government or other person, or (b) use or deployment of weapons incident to U.S. defense activities.[5]

The Nuclear Regulatory Commission (NRC), the AEC's successor regulatory agency, says that the rule "was primarily intended to make clear that privately-owned nuclear facilities were not responsible for defending against attacks that typically could only be carried out by foreign military organizations."[6] NRC's initial DBT, established in the late 1970s, was intended to be consistent with the enemy of the state rule, which remains in effect.

However, the 9/11 attacks drew greater attention to the potential severity of credible terrorist threats. Following the attacks, NRC evaluated the extent to which nuclear plant security forces should be able to defend against such threats, and ordered a strengthening of the DBT, along with other security measures, on April 29, 2003. That order changed the DBT to "represent the largest reasonable threat against which a regulated private guard force should be expected to defend under existing law," according to the NRC announcement.[7]

In the Energy Policy Act of 2005 (EPACT05), Congress imposed a statutory requirement on the NRC to initiate rulemaking for revising the design basis threat.[8] EPACT05 required NRC to consider 12 factors in revising the DBT, such as an assessment of various terrorist threats, sizable explosive devices and modern weapons, attacks by persons with sophisticated knowledge of facility operations, and attacks on spent fuel shipments.

NRC approved its final rule amending the DBT (10 C.F.R. Part 73.1) on January 29, 2007, effective April 18, 2007.[9] Although specific details of the revised DBT were not released to the public, in general the final rule

- clarifies that physical protection systems are required to protect against diversion and theft of fissile material;
- expands the assumed capabilities of adversaries to operate as one or more teams and attack from multiple entry points;

- assumes that adversaries are willing to kill or be killed and are knowledgeable about specific target selection;
- expands the scope of vehicles that licensees must defend against to include water vehicles and land vehicles beyond four-wheel-drive type;
- revises the threat posed by an insider to be more flexible in scope; and
- adds a new mode of attack from adversaries coordinating a vehicle bomb assault with another external assault.

The DBT final rule excluded aircraft attacks as beyond the reasonable responsibility of a private security force, a decision that raised considerable controversy. In approving the rule, NRC rejected a petition from the Union of Concerned Scientists to require that nuclear plants be surrounded by aircraft barriers made of steel beams and cables (the so-called "beamhenge" concept). Critics of NRC's final rule charged that deliberate aircraft crashes were a highly plausible mode of attack, given the events of 9/11. However, NRC contended that power plants were already required to mitigate the effects of aircraft crashes and that "active protection against airborne threats is addressed by other federal organizations, including the military."[10] Additional NRC action on aircraft threats is discussed in the next section.

NRC Commissioners in January 2009 rejected a proposal by the NRC staff to strengthen the classified portion of the DBT to include additional capabilities by potential attackers, according to news reports. The staff proposal lost in a 2-2 vote, with one commissioner position vacant. In an interview afterward, NRC Chairman Dale Klein said the vote could be reconsidered after completion of an ongoing interagency study.[11] Critics contend that the DBT excludes major types of weapons used by terrorists, such as rocket-propelled grenades, and is generally not based on the maximum credible threat identified by the intelligence community.[12]

Critics of NRC's security regulations also have pointed out that licensees are required to employ only a minimum of 10 security personnel on duty per plant, which they argue is not enough for the job.[13] Nuclear spokespersons have responded that the actual security force for the nation's 65 nuclear plant sites numbers more than 5,000, an average of about 75 per site (covering multiple shifts). The industry also points out that nuclear plants all have integrated communications and emergency response plans that include local, state, and federal security forces. The integrated response by outside security forces is intended to handle attacks that might overwhelm an individual plant's security force.[14]

LARGE AIRCRAFT CRASHES

Nuclear power plants were designed to withstand hurricanes, earthquakes, and other extreme events. But deliberate attacks by large airliners loaded with fuel, such as those that crashed into the World Trade Center and Pentagon, were not analyzed when design requirements for today's reactors were determined.[15] Concern about aircraft crashes was intensified by a taped interview shown September 10, 2002, on the Arab TV station al-Jazeera, which contained a statement that Al Qaeda initially planned to include a nuclear plant in its list of 2001 attack sites.

In light of the possibility that an air attack might penetrate the containment structure of a nuclear plant or a spent fuel storage facility, some interest groups have suggested that such an event could be followed by a meltdown or spent fuel fire and widespread radiation exposure. Nuclear industry spokespersons have countered by pointing out that relatively small, low-lying nuclear power plants are difficult targets for attack, and have argued that penetration of the containment is unlikely, and that even if such penetration occurred it probably would not reach the reactor vessel. They suggest that a sustained fire, such as that which melted the steel support structures in the World Trade Center buildings, would be impossible unless an attacking plane penetrated the containment completely, including its fuel-bearing wings. According to former NRC Chairman Nils Diaz, NRC studies, which have not been released, "confirm that the likelihood of both damaging the reactor core and releasing radioactivity that could affect public health and safety is low."[16]

NRC proposed in October 2007 to amend its regulations to require newly designed power reactors to take into account the potential effects of the impact of a large commercial aircraft. [17] As discussed in the previous section, NRC considers an aircraft attack to be beyond the design basis threat that plants must be able to withstand, so the requirements of the proposed rule were intended to provide an additional margin of safety. The proposed rule would affect only new reactor designs not previously certified by NRC, because the previous designs were still considered adequately safe. Nevertheless, Westinghouse submitted changes in the certified design of its AP 1000 reactor to NRC on May 29, 2007, proposing to line the inside and outside of the reactor's concrete containment structure with steel plates to increase resistance to aircraft penetration.[18]

Under NRC's 2007 proposed rule, applicants for new certified designs or for new reactor licenses using uncertified designs would have been required to assess the effects that a large aircraft crash would have on the proposed facilities. Each applicant would then describe how the plant's design features, capabilities, and operations would avoid or mitigate the effects of such a crash, particularly on core cooling, containment integrity, and spent fuel storage pools.

In response to comments, the NRC staff proposed in October 2008 that the aircraft impact assessments be conducted by all new reactors, including those using previously certified designs.[19] The NRC Commissioners, in a 3-1 vote, approved the change February 17, 2009,[20] and it was published in the Federal Register June 12, 2009.[21] The new rule added specific design requirements that all new reactors would have to meet:

> Each applicant subject to this section shall perform a design-specific assessment of the effects on the facility of the impact of a large, commercial aircraft. Using realistic analyses, the applicant shall identify and incorporate into the design those design features and functional capabilities to show that, with reduced use of operator actions:
> (A) the reactor core remains cooled, or the containment remains intact; and
> (B) spent fuel cooling or spent fuel pool integrity is maintained.

As noted above, NRC rejected proposals that existing reactors—in addition to new reactors—be required to protect against aircraft crashes, such as by adding "beamhenge" barriers. NRC determined that damage from aircraft crashes at existing reactors would be sufficiently mitigated by measures that had already been required by all reactors. In 2002, NRC ordered all nuclear power plants to develop strategies to mitigate the effects of large fires and explosions that could result from aircraft crashes or other causes.[22] As part of a broad security rulemaking effort, NRC proposed in October 2006 to incorporate the 2002 order on fire and explosion strategies into its security regulations (10 CFR Part 73).[23] In response to comments, NRC published a supplemental proposed rule in April 2008 to move the fire and explosion requirements into its reactor licensing regulations at 10 CFR Part 50, along with requirements that reactors establish procedures for responding to specific aircraft threat notifications.[24] Those regulations received final approval by the NRC Commissioners December 17, 2008,[25] and were published in the Federal Register March 27, 2009.[26] A key provision in the new rule states:

> Each licensee shall develop and implement guidance and strategies intended to maintain or restore core cooling, containment, and spent fuel pool cooling capabilities under the circumstances associated with loss of

large areas of the plant due to explosions or fire, to include strategies in the following areas:

(i) Fire fighting;

(ii) Operations to mitigate fuel damage; and

(iii) Actions to minimize radiological release.[27]

SPENT FUEL STORAGE

When no longer capable of sustaining a nuclear chain reaction, highly radioactive "spent" nuclear fuel is removed from the reactor and stored in a pool of water in the reactor building and at many sites later transferred to dry casks on the plant grounds. Because both types of storage are located outside the reactor containment structure, particular concern has been raised about the vulnerability of spent fuel to attack by aircraft or other means. If terrorists could breach a spent fuel pool's concrete walls and drain the cooling water, the spent fuel's zirconium cladding could overheat and catch fire.

The National Academy of Sciences (NAS) released a report in April 2005 that found that "successful terrorist attacks on spent fuel pools, though difficult, are possible," and that "if an attack leads to a propagating zirconium cladding fire, it could result in the release of large amounts of radioactive material." NAS recommended that the hottest spent fuel be interspersed with cooler spent fuel to reduce the likelihood of fire, and that water-spray systems be installed to cool spent fuel if pool water were lost. The report also called for NRC to conduct more analysis of the issue and consider earlier movement of spent fuel from pools into dry storage.[28]

NRC agreed with some of findings of the NAS study but disagreed in several areas. In a report to Congress in response to the NAS report, NRC stated:

In summary, the NRC believes based on information developed in NRC vulnerability assessments, that the Committee has identified some scenarios that are unreasonable. The NRC also disagrees with some NAS recommendations and its conclusion lacks a sound technical basis. The NAS finding that earlier movement of spent fuel from pools into dry storage would be prudent is one such example.[29]

NRC conducted the site-specific analyses recommended by NAS with funding provided by the FY2006 Energy and Water Development Appropriations Act (P.L. 109-103, H.Rept. 109-275). NRC's March 2009 regulations cited above include "spent fuel pool cooling capabilities" as a function that must be addressed by nuclear plants' mitigation strategies for large fires and explosions. Protection of spent fuel cooling also is included in the design requirements for new reactors under NRC's June 2009 aircraft impact regulations.

NRC has long contended that the potential effects of terrorist attacks are not "reasonably foreseeable" impacts that must be included in environmental studies for proposed spent fuel storage and other nuclear facilities. However, the U.S. Court of Appeals for the 9th Circuit ruled in June 2006 that terrorist attacks must be included in the environmental study of a dry storage facility at California's Diablo Canyon nuclear plant. NRC reissued the Diablo Canyon study May 29, 2007, to comply with the court ruling, but it did not include terrorism in other recent environmental studies outside the jurisdiction of the 9th Circuit.[30] The U.S. Court of Appeals for the 3rd Circuit subsequently ruled that NRC did not have to consider the impact of terrorist attacks in the license renewal application for the Oyster Creek plant in New Jersey.[31]

Long-term management of spent nuclear fuel is currently undergoing review, but spent fuel stored at reactor sites is expected to be moved eventually to central storage, permanent disposal, or reprocessing facilities. (For details, see CRS Report RL3 346 1, *Civilian Nuclear Waste Disposal*, by Mark Holt.) Large-scale transportation campaigns would increase public attention to NRC transportation security requirements and related security issues.

FORCE-ON-FORCE EXERCISES

EPACT05 codified an NRC requirement that each nuclear power plant conduct security exercises every three years to test its ability to defend against the design basis threat. In these "force-onforce" exercises, closely monitored and evaluated by NRC, a mock adversary force from outside the plant attempts to penetrate the plant's vital area and simulate damage to a "target set" of key safety components. Actual damage to such components could result in radioactive releases from the plant. Participants in the tightly controlled exercises carry weapons modified to fire only blanks and laser bursts to simulate bullets, and they wear laser sensors to indicate hits. Other weapons and explosives, as well as destruction or breaching of physical security barriers, may also be simulated. While one squad of the plant's guard force is participating in a force-on-force exercise, another squad is also on duty to maintain normal plant security. Plant defenders know that a mock attack will take place sometime during a specific period of several hours, but they do not know what the attack scenario will be. Multiple attack scenarios are conducted over several days of exercises.

Full implementation of the force-on-force program began in late 2004. Standard procedures and other requirements have been developed for using the force-on-force exercises to evaluate plant security and as a basis for taking regulatory enforcement action. Many tradeoffs are necessary to make the exercises as realistic and consistent as possible without endangering participants or regular plant operations and security.

NRC required the nuclear industry to develop and train, under NRC standards, a "composite adversary force" comprising security officers from many plants to simulate terrorist attacks in all force-on-force exercises conducted after October 2004. However, in September 2004 testimony, GAO criticized the industry's selection of Wackenhut (now G4S Regulated Security Solutions), a security company that guards many U.S. nuclear plants, to manage the adversary force, including non-Wackenhut employees. In addition to raising "questions about the force's independence," GAO noted that Wackenhut had been accused of cheating on previous force-on-force exercises by the Department of Energy.[32] Exelon terminated its security contracts with Wackenhut in late 2007 after guards at the Peach Bottom reactor in York County, Pennsylvania, were discovered sleeping while on duty. [33]

EPACT05 requires NRC to "mitigate any potential conflict of interest that could influence the results of a force-on-force exercise, as the Commission determines to be necessary and appropriate." NRC prohibits officers in the adversary force from participating in exercises at their home plants. As in previous years, NRC's 2009 annual security report to Congress found that the industry adversary teams "continued to meet expectations for a credible, well-trained, and consistent mock adversary force."[34]

Through calendar year 2009, NRC had completed two-thirds of the second three-year cycle of force-on-force exercises at the 64 U.S. nuclear plant sites.[35] From the start of the program in 2004 through 2009, 112 force-on-force inspections were conducted, with each inspection typically including three mock attacks by the adversary force. During the 112 inspections, 8 complete target sets were simulated to be damaged or destroyed, indicating inadequate protection against the DBT, and additional security measures were promptly implemented, according to NRC. The inspections resulted in a total of 49 findings of security deficiencies, 40 of which were of relatively low significance. Follow-up force-on-force exercises are sometimes conducted to verify that the necessary security improvements have been made. [36]

EMERGENCY RESPONSE

After the 1979 accident at the Three Mile Island nuclear plant near Harrisburg, PA, Congress required that all nuclear power plants be covered by emergency plans. NRC requires each plant to have an Emergency Planning Zone (EPZ) with an approximately 10-mile radius. Within the emergency EPZ, the plant operator must maintain warning sirens or other systems and regularly conduct emergency response exercises evaluated by NRC and the Federal Emergency Management Agency (FEMA).

In light of the increased possibility of terrorist attacks that, if successful, could result in the release of radioactive material, proposals have been made to expand the EPZ to include larger population centers. NRC determined that the 10-mile EPZ remained adequate, but it issued a bulletin in July 2005 identifying enhancements for emergency response plans in the case of "security-based events at a nuclear power plant."[37]

The potential release of radioactive iodine during a nuclear incident is a particular concern, because iodine tends to concentrate in the thyroid gland of persons exposed to it. Emergency plans in many states include distribution of iodine pills to the population within the EPZ. Taking non-radioactive iodine before exposure would prevent absorption of radioactive iodine but would afford no protection against other radioactive elements. In 2002, NRC began providing iodine pills to states requesting them for populations within the 10-mile EPZ.

SECURITY PERSONNEL AND OTHER ISSUES

After video recordings of inattentive security officers at the Peach Bottom (PA) nuclear power plant were aired on local television, an NRC inspection in late September 2007 confirmed that there had been multiple occasions on which multiple security officers were inattentive.[38] However, after a follow-up inspection into security issues at the Peach Bottom plant, run by Exelon Nuclear, the NRC concluded that the plant's security program had not been significantly degraded as a result of the guards' inattentiveness. NRC issued a bulletin December 12, 2007, requiring all nuclear power plants to provide written descriptions of their "managerial controls to deter and address inattentiveness and complicity among licensee security personnel."[39]

The incident drew harsh criticism from the House Committee on Energy and Commerce. "The NRC's stunning failure to act on credible allegations of sleeping security guards, coupled with its unwillingness to protect the whistleblower who uncovered the problem, raises troubling questions," said Representative John D. Dingell, then-Chairman of the Committee.[40] NRC proposed a $65,000 fine on Exelon Nuclear on January 6, 2009.[41]

Following the 9/11 terrorist attacks, NRC conducted a "top-to-bottom" review of its nuclear power plant security requirements. On February 25, 2002, the agency issued "interim compensatory security measures" to deal with the "generalized high-level threat environment" that continued to exist, and on January 7, 2003, it issued regulatory orders that tightened nuclear plant access. On April 29, 2003, NRC issued orders to restrict security officer work hours, establish new security force training and qualification requirements, and increase the DBT that nuclear security forces must be able to defend against, as discussed previously.

In October 2006, NRC proposed to amend the security regulations and add new security requirements that would codify the series of orders issued after 9/11 and respond to requirements in the Energy Policy Act of 2005.[42] The new security regulations were approved by the NRC Commissioners on December 17, 2008, and published March 27, 2009[43]:

- *Safety and Security Interface.* Explicit requirements are established for nuclear plants to ensure that necessary security measures do not compromise plant safety.
- *Mixed-Oxide Fuel.* Enhanced physical security requirements are established to prevent theft or diversion of plutonium-bearing mixed-oxide (MOX) fuel.
- *Cyber Security.* Nuclear plants must submit security plans that describe how digital computer and communications systems and safety-related networks are protected from cyber attacks.
- *Aircraft Attack Mitigative Strategies and Response.* As discussed in the earlier section on vulnerability to aircraft crashes, nuclear plants must prepare strategies for responding to warnings of an aircraft attack and for mitigating the effects of large explosions and fires.

- *Plant Access Authorization.* Nuclear plants must implement more rigorous programs for authorizing access, including enhanced psychological assessments and behavioral observation.
- *Security Personnel Training and Qualification.* Modifications to security personnel requirements include additional physical fitness standards, increased minimum qualification scores for mandatory personnel tests, and requirements for on-the-job training.
- *Physical Security Enhancements.* New requirements are intended to ensure the availability of backup security command centers, uninterruptible power supplies to detection systems, enhanced video capability, and protection from waterborne vehicles.

A proposal by NRC staff to release more details about the results of nuclear plant security inspections was defeated by the NRC Commissioners in a 2-2 vote on January 21, 2009. Under current policy, NRC announces after a security inspection whether any violations that were found were of low safety significance or moderate-or-higher safety significance. Critics of the current policy contend that the public needs more detail to be assured of plant security. The policy's supporters counter that greater information about security inspection findings could inadvertently provide useful information to terrorists.[44]

End Notes

[1] Homeland Security Presidential Directive 7: Critical Infrastructure Identification, Prioritization, and Protection, December 17, 2003, http://www.dhs.gov/xabout/laws
[2] General NRC requirements for nuclear power plant security can be found in 10 C.F.R. 73.55.
[3] 10 C.F.R. § 73.1.
[4] It was feared that Cuba might launch an attack on Florida reactors. Government Accountability Office, *Nuclear Power Plants—Efforts Made to Upgrade Security, but the Nuclear Regulatory Commission's Design Basis Threat Process Should Be Improved* (GAO-06-388), March 2006, p. 2. Regulations at 10 CFR 50.13.
[5] 10 C.F.R. § 50.13. Attacks and destructive acts by enemies of the United States; and defense activities.
[6] Nuclear Regulatory Commission, "Design Basis Threat," 72 *Federal Register* 12714, March 19, 2007.
[7] *Federal Register*, May 7, 2003 (vol. 68, no. 88). NRC, All Operating Power Reactor Licensees; Order Modifying Licenses.
[8] P.L. 109-58, Title VI, Subtitle D—Nuclear Security (Secs. 651-657). Sec. 651 adds Atomic Energy Act Sec. 170E. Design Basis Threat Rulemaking.
[9] *Federal Register*, March 19, 2007 (vol. 72, no. 52), NRC, Design Basis Threat, Final Rule, pp. 12705-12727.

[10] NRC, "NRC Approves Final Rule Amending Security Requirements," News Release No. 07-0 12, January 29, 2007.

[11] Jeff Beattie, "NRC Chairman Questions Case for Tougher DBT," *Energy Daily*, February 17, 2009, p. 1.

[12] Edwin S. Lyman, "Security Since September 11th," *Nuclear Engineering International*, March 2010, pp. 14-19.

[13] 10 C.F.R. 73.55 (k)(5)(ii) states: "The number of armed responders shall not be less than ten (10)." The previous requirement, in 10 C.F.R. 73.55 (h)(3), stated: "The total number of guards, and armed, trained personnel immediately available at the facility to fulfill these response requirements shall nominally be ten (10), unless specifically required otherwise on a case by case basis by the Commission; however, this number may not be reduced to less than five (5) guards." The change was made in NRC final regulations published in March 2009, op. cit.

[14] Doug Walters, "Security Since March," *Nuclear Engineering International*, May 2010.

[15] Meserve, Richard A., NRC Chairman, "Research: Strengthening the Foundation of the Nuclear Industry," Speech to Nuclear Safety Research Conference, October 29, 2002.

[16] Letter from NRC Chairman Nils J. Diaz to Secretary of Homeland Security Tom Ridge, September 8, 2004.

[17] *Federal Register*, October 3, 2007 (vol. 72, no. 191), Consideration of Aircraft Impacts for New Nuclear Power Reactor Designs.

[18] MacLachlan, Ann, "Westinghouse Changes AP1000 Design to Improve Plane Crash Resistance," Nucleonics Week, June 21, 2007.

[19] Nuclear Regulatory Commission, *Final Rule—Consideration of Aircraft Impacts for New Nuclear Power Reactors*, Rulemaking Issue Affirmation, SECY-08-0152, October 15, 2008.

[20] Nuclear Regulatory Commission, *Final Rule—Consideration of Aircraft Impacts for New Nuclear Power Reactors*, Commission Voting Record, SECY-08-0 152, February 17, 2009.

[21] Nuclear Regulatory Commission, Consideration of Aircraft Impacts for New Nuclear Power Reactors, Final Rule, 74 *Federal Register* 28111, June 12, 2009. This provision is codified at 10 CFR 50.150.

[22] Nuclear Regulatory Commission, *Final Rule—Consideration of Aircraft Impacts for New Nuclear Power Reactors*, Rulemaking Issue Affirmation, SECY-08-0152, October 15, 2008, p. 2.

[23] Nuclear Regulatory Commission, "Power Reactor Security Requirements, Proposed Rule," 71 *Federal Register* 62664, October 26, 2006.

[24] Nuclear Regulatory Commission, "Power Reactor Security Requirements, Supplemental Proposed Rule," 73 *Federal Register* 19443, April 10, 2008.

[25] Nuclear Regulatory Commission, "NRC Approves Final Rule Expanding Security Requirements for Nuclear Power Plants," press release, December 17, 2008, http://www.nrc.gov/reading

[26] Nuclear Regulatory Commission, Power Reactor Security Requirements, Final Rule, 74 *Federal Register* 13925, March 27, 2009.

[27] 10 CFR 50.54(hh)(2).

[28] National Academy of Sciences, Board on Radioactive Waste Management, Safety and Security of Commercial Spent Nuclear Fuel Storage, Public Report (online version), released April 6, 2005.

[29] U.S. Nuclear Regulatory Commission Report to Congress on the National Academy of Sciences Study on the Safety and Security of Commercial Spent Nuclear Fuel Storage, March 2005, p. iii, http://www.nrc.gov/reading- 142005 .pdf.

[30] Beattie, Jeff, "NRC Takes Two Roads on Terror Review Issue," Energy Daily, February 27, 2007.

[31] U.S. Court of Appeals for the Third Circuit, *New Jersey Department of Environmental Protection v. U.S. Nuclear Regulatory Commission*, March 31, 2009, http://www.ca3.uscourts.gov/opinarch/072271p.pdf.

[32] GAO. "Nuclear Regulatory Commission: Preliminary Observations on Efforts to Improve Security at Nuclear Power Plants." Statement of Jim Wells, Director, Natural Resources and Environment to the Subcommittee on National Security, Emerging Threats, and International Relations, House Committee on Government Reform. September 14, 2004. p. 14.

[33] *Washington Post*, "Executive Resigns in Storm Over Sleeping Guards," January 10, 2008.

[34] Nuclear Regulatory Commission, Office of Nuclear Security and Incident Response, Report to Congress on the Security Inspection Program for Commercial Power Reactor and Category 1 Fuel Cycle Facilities: Results and Status Update; Annual Report for Calendar Year 2009, NUREG-1885, Rev. 3, July 2010, p. 8, http://www.nrc.gov/readingrm/doc-collections/nuregs/staff/sr1 885/r3/sr1 885r3.pdf.

[35] NRC generally lists 65 U.S. plant sites, but the adjacent Hope Creek and Salem sites in New Jersey are considered to be a single site for security exercises. E-mail message from David Decker, NRC Office of Congressional Affairs, March 13, 2009.

[36] NRC Office of Nuclear Security and Incident Response, op. cit.

[37] NRC Office of Nuclear Security and Incident Response, *Emergency Preparedness and Response Actions for Security-Based Events*, NRC Bulletin 2005-02, July 18, 2005, http://www.nrc.gov/reading-

[38] NRC, *NRC Commences Follow-up Security Inspection at Peach Bottom*, November 5, 2007, http://www.nrc.gov/ reading-rm/doc-collections/news/2007/07-057.i.html.

[39] Nuclear Regulatory Commission, *Security Officer Attentiveness*, NRC Bulletin 2007-1, Washington, DC, December 12, 2007.

[40] Committee on Energy and Commerce, *Energy and Commerce Committee to Probe Breakdowns in NRC Oversight*, January 7, 2008 http://energycommerce.house.gov/ Press_110/110nr149.shtml.

[41] Nuclear Regulatory Commission, "NRC Proposes $65,000 Fine for Violations Associated with Inattentive Security Guards at Peach Bottom Nuclear Plant," press release, January 6, 2009, http://www.nrc.gov/reading-

[42] *Federal Register*, October 26, 2006 (vol. 71, no. 207), NRC, Power Reactor Security Requirements, Proposed Rule.

[43] *Federal Register*, March 27, 2009, op. cit.

[44] Jenny Weil, "Commissioners Reach Stalemate on Security-Related Amendment," *Inside NRC*, February 2, 2009.

In: Nuclear Power Plants   ISBN: 978-1-61470-952-7
Editor: James P. Argyriou   ©2012 Nova Science Publishers, Inc.

Chapter 6

NUCLEAR INSURANCE AND DISASTER RELIEF FUNDS

U.S. Nuclear Regulatory Commission

NUCLEAR INSURANCE: PRICE-ANDERSON ACT

The Price-Anderson Act, which became law on September 2, 1957, was designed to ensure that adequate funds would be available to satisfy liability claims of members of the public for personal injury and property damage in the event of a nuclear accident involving a commercial nuclear power plant. The legislation helped encourage private investment in commercial nuclear power by placing a cap, or ceiling on the total amount of liability each holder of a nuclear power plant licensee faced in the event of an accident. Over the years, the "limit of liability" for a nuclear accident has increased the insurance pool to more than $12 billion.

Under existing policy, owners of nuclear power plants pay a premium each year for $375 million in private insurance for offsite liability coverage for each reactor unit. This primary, or first tier, insurance is supplemented by a second tier. In the event a nuclear accident causes damages in excess of $375 million, each licensee would be assessed a prorated share of the excess up to $111.9 million. With 104 reactors currently licensed to operate, this secondary tier of funds contains about $12.6 billion. If 15 percent of these funds are expended, prioritization of the remaining amount would be left to a federal district court. If the second tier is depleted, Congress is committed to determine whether additional disaster relief is required.

The only insurance pool writing nuclear insurance, American Nuclear Insurers, is comprised of investor- owned stock insurance companies. About half the pool's total liability capacity comes from foreign sources. The average annual premium for a single-unit reactor site is $400,000. The premium for a second or third reactor at the same site is discounted to reflect a sharing of limits.

Claims resulting from nuclear accidents are covered under Price-Anderson; for that reason, all property and liability insurance policies issued in the U.S. exclude nuclear accidents. Claims can include any incident (including those that come about because of theft or sabotage) in the course of transporting nuclear fuel to a reactor site; in the storage of nuclear fuel or waste at a site; in the operation of a reactor, including the discharge of radioactive effluent; and in the transportation of irradiated nuclear fuel and nuclear waste from the reactor.

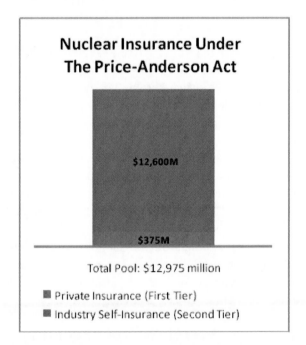

Owners of nuclear power plants pay for $375 million private insurance. If a nuclear accident surpasses this amount, each plant pays up to $111.8 million into a second tier insurance pool.

Price-Anderson does not require coverage for spent fuel or nuclear waste stored at interim storage facilities, transportation of nuclear fuel or waste that is not either to or from a nuclear reactor, or acts of theft or sabotage occurring after planned transportation has ended.

Insurance under Price-Anderson covers bodily injury, sickness, disease or resulting death, property damage and loss as well as reasonable living expenses for individuals evacuated.

The Energy Policy Act of 2005 extended the Price-Anderson Act to December 31, 2025.

PRICE-ANDERSON IN ACTION

When the accident at Three Mile Island Nuclear Power Plant in Middletown, Pa., occurred in 1979, the Price-Anderson Act provided liability insurance to the public. Coverage was available to those in need by the time Pennsylvania's governor recommended the evacuation of pregnant women and families with young children who lived near the plant. At the time of the accident, private insurers had $140 million of coverage available in the first tier pools. Insurance adjusters advanced money to evacuated families in order to cover their living expenses, only requesting that unused funds be returned; recipients responded by sending back several thousand dollars. The insurance pools also reimbursed over 600 individuals and families for wages lost as a result of the accident.

In addition to the immediate concerns, the insurance pools were later used to settle a class-action suit for economic loss filed on behalf of residents who lived near Three Mile Island. Because the Price-Anderson Act allowed for a certain amount of money to be spent on each accident, it covered court fees as well. The last of the litigation surrounding the accident was resolved in 2003.

To date, the insurance pools have paid approximately $71 million in claims and litigation costs associated with the Three Mile Island accident.

DISASTER RELIEF FUNDS: STAFFORD ACT

Disaster relief is also available to State and local governments under the Robert T. Stafford Disaster Relief and Emergency Assistance Act if a nuclear accident is declared an emergency or major disaster by the President. The Act is designed to provide early assistance to accident victims. Under a cost-sharing provision, State governments pay 25 percent of the cost of temporary housing for up to 18 months, home repair, temporary mortgage or rental payments and other "unmet needs" of disaster victims; the federal government pays the balance.

CHAPTER SOURCES

The following chapters have been previously published:

Chapter 1 – This is an edited reformatted and augmented version of a Congressional Research Service publication, report R41805, dated May 2, 2011.

Chapter 2 – This is an edited reformatted and augmented version of a United States Nuclear Regulatory Commission publication, dated April 2011.

Chapter 3 – This is an edited reformatted and augmented version of a United States Nuclear Regulatory Commission publication, dated June 2008.

Chapter 4 – This is an edited reformatted and augmented version of a United States Department of Commerce publication, dated February 2011.

Chapter 5 - This is an edited, reformatted and augmented version of Congressional Research Service publication, Report RL34331, dated August 23, 2010.

Chapter 6 – This is an edited reformatted and augmented version of a United States Nuclear Regulatory Commission publication, dated August 2010.

INDEX

D

E